AF777032

Recent Advances in Boron-Containing Materials

Edited by Metin Aydin

Published in London, United Kingdom

IntechOpen

Supporting open minds since 2005

Recent Advances in Boron-Containing Materials
http://dx.doi.org/10.5772/intechopen.78730
Edited by Metin Aydin

Contributors
Sasikumar Sasikumar Rathinasabapathy, A Manivannan, Helio Ribeiro, Paula Von Randow C, Diego N. Vilela, Lidia M. Andrade, Milene Adriane Luciano, Pavel Stuzhin, Georgy Pakhomov, Vlad Travkin, Ingrid Garces Millas, Anthony Harriman, Hatun Al-Sharif, Maria Prokopivna Savyak

© The Editor(s) and the Author(s) 2020
The rights of the editor(s) and the author(s) have been asserted in accordance with the Copyright, Designs and Patents Act 1988. All rights to the book as a whole are reserved by INTECHOPEN LIMITED. The book as a whole (compilation) cannot be reproduced, distributed or used for commercial or non-commercial purposes without INTECHOPEN LIMITED's written permission. Enquiries concerning the use of the book should be directed to INTECHOPEN LIMITED rights and permissions department (permissions@intechopen.com).
Violations are liable to prosecution under the governing Copyright Law.

Individual chapters of this publication are distributed under the terms of the Creative Commons Attribution 3.0 Unported License which permits commercial use, distribution and reproduction of the individual chapters, provided the original author(s) and source publication are appropriately acknowledged. If so indicated, certain images may not be included under the Creative Commons license. In such cases users will need to obtain permission from the license holder to reproduce the material. More details and guidelines concerning content reuse and adaptation can be found at http://www.intechopen.com/copyright-policy.html.

Notice
Statements and opinions expressed in the chapters are these of the individual contributors and not necessarily those of the editors or publisher. No responsibility is accepted for the accuracy of information contained in the published chapters. The publisher assumes no responsibility for any damage or injury to persons or property arising out of the use of any materials, instructions, methods or ideas contained in the book.

First published in London, United Kingdom, 2020 by IntechOpen
IntechOpen is the global imprint of INTECHOPEN LIMITED, registered in England and Wales, registration number: 11086078, 7th floor, 10 Lower Thames Street, London, EC3R 6AF, United Kingdom
Printed in Croatia

British Library Cataloguing-in-Publication Data
A catalogue record for this book is available from the British Library

Additional hard and PDF copies can be obtained from orders@intechopen.com

Recent Advances in Boron-Containing Materials
Edited by Metin Aydin
p. cm.
Print ISBN 978-1-83880-040-6
Online ISBN 978-1-83880-213-4
eBook (PDF) ISBN 978-1-83880-214-1

For EU product safety concerns:
IN TECH d.o.o., Prolaz Marije Krucifikse Kozulić 3, 51000 Rijeka, Croatia, info@intechopen.com or visit our website at intechopen.com.

We are IntechOpen, the world's leading publisher of Open Access books

Built by scientists, for scientists

4,800+
Open access books available

122,000+
International authors and editors

135M+
Downloads

151
Countries delivered to

Our authors are among the
Top 1%
most cited scientists

12.2%
Contributors from top 500 universities

WEB OF SCIENCE™

Selection of our books indexed in the Book Citation Index in Web of Science™ Core Collection (BKCI)

Interested in publishing with us?
Contact book.department@intechopen.com

Numbers displayed above are based on latest data collected.
For more information visit www.intechopen.com

Meet the editor

Dr. Aydin received his BS degree in 1989 and his MS degree in Engineering Physics in 1993 from Ankara University. He also holds a doctorate in Physical Chemistry from the City University of New York (CUNY), received in 2001. From 2001 to 2005, Dr. Aydin was a post-doctoral associate at the Center for Analysis of Structures and Interfaces, The City College, Chemistry Department, at CUNY. In 2008, Dr. Aydin joined the faculty of the Chemistry Department of Ondokuz Mayıs University, where he rose through the ranks from assistant professor to professor of physical chemistry. Dr. Aydin's work has been published in numerous professional journals. He has talked extensively at conferences and universities, both national and international. His main research interests involve photofragmentation spectroscopy of transition metal cation clusters in a reflectron time-of-flight mass spectrometer, surface enhanced Raman spectroscopy, aggregation enhanced Raman spectroscopy, density functional theory, chemistry and physics of isolated and functionalized carbon nanotubes, boron nitride nanotubes and graphene, and X-ray diffraction.

Contents

Preface XI

Chapter 1 1
Boron Industry, Sources, and Evaporitic Andean Deposits: Geochemical Characteristics and Evolution Paths of the Superficial Brines
by Ingrid Garcés Millas

Chapter 2 25
Functionalized Boron Nitride Applications in Biotechnology
by Hélio Ribeiro, Paula von Cardoso Randow, Diego N. Vilela, Milene Adriane Luciano and Lidia Maria de Andrade

Chapter 3 45
Bio-Inspired Artificial Light-Harvesting Arrays Based on Boron(III)-Chelates
by Hatun H.T. Al-Sharif and Anthony Harriman

Chapter 4 63
Significance of Boron Nitride in Composites and Its Applications
by Sasikumar Rathinasabapathy, M.S. Santhosh and Manivannan Asokan

Chapter 5 75
Hexachlorinated Boron(III) Subphthalocyanine as Acceptor for Organic Photovoltaics: A Brief Overview
by Georgy L. Pakhomov, Vlad V. Travkin and Pavel A. Stuzhin

Chapter 6 93
Peculiarities of Refractory Borides Formation during Mechanical Alloying IV-V Group Transition Metals with Boron in Planetary Mill
by Maria P. Savyak and Alex B. Melnick

Preface

Boron is a semimetal that is found in low amounts both in the solar system and in the Earth's crust. It is a chemical element with atomic number 5 and is produced by cosmic ray and high-energy radiation from supernovae that impact boron-containing minerals. Naturally occurring boron is found within chemical compounds such as borate minerals; however, the elemental form is not found because of its high reactivity. Also, high purity boron is difficult to produce industrially because of contamination by carbon and other elements at high temperatures. Boron has several allotropes, including amorphous boron, a dark brown powder, and crystalline boron, a black material with a melting point above 2000°C, which is extremely hard and weakly conductive at room temperature, but a good conductor at high temperature.

Boron is an important synthetic and research chemical. It is an important component of reducing agents such as sodium borohydride and borane. In the Corey–Bakshi–Shitaba catalyst used to asymmetrically reduce ketones, boron plays dual roles as a hydride source and a Lewis acid. Boron's Lewis acid properties result from an empty p-orbital, which confers electron-accepting abilities, such as in the familiar boron trifluoride used as a Lewis acid catalyst. Boronic acids and esters are key organic building blocks in synthetic chemistry: these oxygen- and carbon-containing boron compounds are important cross-coupling partners in palladium-catalyzed Suzuki–Miyaura reactions. They are also critical to the small molecule synthesis of pharmaceuticals, agrochemicals, and veterinary science products.

Recent developments in the field of boron chemistry have led to a broad range of applications from medical science to materials uses. Boron-doped drugs have several biomedical applications, including use in neutron capture therapy for cancer, as an anticancer, antiviral, antibacterial, antifungal therapeutic agent, and as a component of optical and nuclear imaging agents. Due to the electron-deficient nature of the boron atom, boron-based organic materials, such as three-coordinate organoboron compounds, present electron-accepting centers that give exceptional optoelectronic functions and significantly amplify performance in energy-conversion systems such as organic light-emitting diodes, organic photovoltaics, and organic field-effect transistors. Furthermore, the rapid development of nanoscience and nanotechnology has inspired scientists to investigate new features of boron nanostructures at the nanoscale such as boron-containing nanotubes, nanowires, and nanosheets. These novel structures have found applications in the field of hydrogen storage, batteries, catalysts, electronics, superconductors, mechanically strong components, nanomedicine, and cancer research.

The aim of this book is to present an overview of recent developments in boron-containing materials with potential applications in many fields. In Chapter 1, the history of boron and its usage areas are briefly reviewed, including a comprehensive discussion on boron deposits and sources in Chile. Chapter 2 provides experimental and theoretical works for different boron nitride functionalization structures with potential biomedical applications. Chapter 3 deals with the photophysical and photochemical properties of boron (III) chelates for photonic and sunlight harvesting applications. Chapter 4 deals with properties and applications of boron nitride nanomaterials. Chapter 5 discusses experimental results in hexachlorinated boron (III) subphthalocyanine as an acceptor for organic photovoltaics. Experimental and theoretical studies of mechanical alloying in a transition metal (IV–V group)-boron system are presented in Chapter 6.

Metin Aydin, PhD
Professor of Chemistry,
Ondokuz Mayıs University,
Samsun, Turkey

Chapter 1

Boron Industry, Sources, and Evaporitic Andean Deposits: Geochemical Characteristics and Evolution Paths of the Superficial Brines

Ingrid Garcés Millas

Abstract

This study accounts for boron deposits in Chile. In addition, a vision is given of the geochemical evolution of its waters that depend largely on the evaporation of water and various factors among which are the geomorphology, climate, and volcanic activity that favor the conditions for the deposition of various salts between that accentuate lithium, potassium, and boron. Borates are found in lenticular stratified bodies, known as "bars," interspersed in detrital-saline sequences and always in the first meters of the saline surface part or as high-grade nodules (up to 30% B_2O_3) that can reach tens of centimeters. In the first part, a description is made of the saline deposits to coming of the salars of South America, because it is the most important reserves of boron-rich minerals known in this continent are directly related to this type of deposits. Subsequently, the deposits in Chile and their characteristics are described. The only mineral of economic recovery known in Chile is ulexite. The second part refers to the Pitzer ion interaction model that is applied to predict the precipitation of salts in multicomponent aqueous systems with high ionic strength in a temperature range of 0–60°C, using the three natural brines of Andean borates.

Keywords: deposits of borates, Andean salars, natural brines, Pitzer model, Salar de Quisquiro, Salar de Aguas Calientes, Salar de Surire

1. Introduction

The boron element does not exist by itself in nature; it appears in combination with oxygen and other elements in salts, commonly called "borates" and defined as a compound that contains boric oxide (B_2O_3). In Babylon, more than 4000 years ago, the use of borax was known because the Babylonians brought it from the Himalayas to use in the manufacture of jewelry. The Egyptians used borax for the mummification process, and by ~300 d.C. the Chinese were familiar with borax glazes, as were the Arabs three centuries later [1–3]. More than 230 minerals containing boron have been identified [2], with sodium, calcium, or magnesium salts being the most common. There are many minerals that contain boric oxide, but four are the most important minerals from the commercial point of view: borax, kernite, ulexite, and colemanite. All are derived from continental evaporites that form the main commercial sources of borates. Deposits that contain these minerals are mined in a limited number of countries (**Figure 1**), dominated by the United States and Turkey, which together furnish 90% of the world's borate supplies, but the most important world reserves are in Turkey (borax, ulexite, and colemanite).

The main uses of borates and boron compounds are found in the enamel and ceramic industry (sanitary ware, crockery, ceramics, tiles), glass industry (crystals, glass fibers, fire-resistant Pyrex glasses, lamps and spotlights), and fertilizers. Kernite is used to produce boric acid, tincal is used to produce sodium borate, and ulexite is used as the principal ingredient in the manufacture of a variety of specialty glasses and ceramics. Boron is one of the seven essential micronutrients for plants; it is applied directly to the soil and also is utilized in the fertilizer-type specialty [4]. It is present in the chemical industry (preparation of detergents, bleach, fire retardants, abrasives, cosmetics), tanneries (prevents rot), pharmacy (mild antiseptic), and paints (fungicide) and is used as a wood preservative (because of its low toxicity) and in capacitors, alloys, catalysis, rubber (fire retardant), and cement (slows setting). Boron hydrides oxidize easily and release large amounts of energy and for that are being studied as a possible source of fuel (aircraft and rocket borane); also they are used for shielding against radiation and for the detection of neutrons. B-10 is used in the control of nuclear reactors, in optics, and in the production of semiconductors [5].

The majority of commercial borate deposits in the world are extracted by open-pit methods. Another main source is from saline lakes, whose deposits are associated with the volcanism of the Neogene age in tectonically active regions on plate boundaries with arid climates, such as in the Mojave Desert of the United States near Boron, California, the Tethyan belt in West Asia, and the Andean belt in South America. The most important continental borates are found in the United States,

Figure 1.
Places where borate deposits are found in the world. The South American borates are located in the 70° W latitude strip and around the 30° N longitude. Turkish borates, American borates, and Chinese Asians are located.

Argentina, Chile, Peru, and China. The world's largest known borate reserves are found in western Anatolia, Turkey [6].

The presence of borates in South America is interesting. Peru has only one site (Laguna Salinas, Arequipa) with reserves that exceed 10 million tons with a 25% law of boric anhydride (B_2O_3) [7]. Bolivia has reserves of ulexite in Coipasa, Empexa, Uyuni, Chiguana, Pastos Grandes, Capina, Mamacoma, Curuto, Chalviri, Luriques, and others, while Argentina has the borate salt flats of Hombre Muerto, Diablillos, Ratones, Centenario, Cauchari, Olaroz, and Salinas Grandes [7].

There are only four major important metallogenic provinces in the world of continental borates. In order of importance are Anatolia in Turkey; Nevada, California, United States; the Central Andes of South America; and Tibet in Central Asia [8] (**Figure 1**). The Andean Boratífera Province includes a part of the south of Peru, the central-eastern part of the Great North of Chile, the high plateau sectors of the northwest of Argentina, and the southwest of Bolivia that has been described by Alonso [9], corresponding to the region between 16 and 27° S and 66 and 70° W with its north–south major axis of the order of 1500 km and another east–west of 400 km. Argentina has the best reserves and production in South America [9].

The Central Andean volcanic province coincides with the normal subduction that accompanies the Altiplano-Puna plateau that has the anomaly in boron and other volcanogenic elements, such as lithium and arsenic. The Andean borate formation model was developed by argentine investigators [9–11]. The concentration of boron that gave place to the formation of deposits of economic interested is related with the concurrence of several associated factors such as volcanism; presence of closed basins; semiarid climate; and thermal sources [7] of the Miocene and Quaternary.

The world production of borates is estimated at 5.75 million tons of different minerals [12], and the main producers are Turkey (2.50 mill ton), United States (1.15 mill ton), Argentina (600 mil ton), Chile (560 mil ton), Russia (400 mil ton), Peru (290 mil ton), and others (270 mil ton for the rest of the producers) [13]. The United States and Turkey are the largest producers of boron in the world. Boron is quoted and sold based on boron oxide (B_2O_3), which varies according to the mineral and the compound and with or without sodium and calcium.

In South America, the "Central Andean Boratífera Province" is located between the Cordillera de la Costa in Chile and the eastern edge of the Altiplano-Puna. The current production of Bolivia is based on the mineral reserves of boron from saline deposits in the south of the country and from the Salar de Uyuni. The reserves in these deposits have been estimated at 15 million tons (Mt) of B_2O_3. The most important boron ore that is extracted is ulexite, associated with tincal [13]. In Peru, there is only one company that extracts natural borates from the Laguna Salinas deposit, where reserves are estimated at 10 Mt. of ore with 25–27% of B_2O_3. The open-pit mine is about 80 km away from Arequipa at an altitude of 4100 m.a.s.l. The capacity of the mine is 120,000 tpa of ulexite and 8000 tpa of colemanite. There are two main types of borates in Puna, Argentina: borates in rocks and borates in salars. The borate deposits in rock are of the Tertiary and Quaternary, formed in the last 7 million years, being located with greater abundance and economic importance in the NOA region (Argentine Northwest). The main mineralogy that these deposits present are borates of sodium (tincal), calcium and sodium (ulexite), calcium (colemanite), and calcium and magnesium (hydroboracite). The reserves in the Tertiary deposits (hard borates) amount to 60 million tons of B_2O_3 and are represented by Tincalayu (Salar del Hombre Muerto, Salta), Sijes (Pastos Grandes, Salta), and Loma Blanca (Puna Jujeña), while the reserves in the Quaternary deposits (soft borates) amount to 40 million tons of B_2O_3 and represent salt flats of Salta, Jujuy, and Catamarca, whose main mineral is ulexite. The brine reserves of borates from salars of La Puna are not all estimated [14].

Chile has been the main producer of borate in South America, and the production in 2017 reached 560,000 tons of ulexite; the largest ulexite salt deposit in the world is the Salar de Surire, with estimated reserves of 1.5 Mt. [12]. Chile has ulexite reserves that together exceed 35 million tons with a 25% grade in B_2O_3, while world reserves are estimated at 380 million tons in B_2O_3 [12]. Canada, China, India, Japan, and Malaysia are the countries that imported the highest amounts of refined borates from the United States in 2018.

2. Review of boron deposits in Chile

The discovery and commercial development of borate deposits were accelerated in the nineteenth century. Chile started to mine borate from the Salar de Ascotán in 1852 (accounting for a quarter of the world's annual supply of ~16,000 tons) by the British company Borax Consolidated [15]. It is necessary to consider that at that time, the borate deposits were not part of the Chilean territory. Chile was the world's principal producer of boron, with its deposits located in the salars of Surire in the north of the country; Ascotán, Carcote, Aguas Calientes I, and Quisquiro, located in the second region; and Maricunga in the third region of Atacama. In 1836 the borates of the South American region were known, with existing records of the year 1852. The incipient exploitation of the Salar de Ascotán in 1883 was described with a production of 36,000 metric tons in 1913 and stopped working in 1967 [16, 17]. Currently, Chile is located after Argentina in reserves and installed treatment capacity among the Andean producing countries. Boron (ulexite) production in Chile increased by approximately 6%, from 518 tons in 2015 to 560 tons in 2017 [18].

The discovery of deposits of borates "hard" in Turkey of minerals with better economic performance, greater reserves, and comparative advantages of exploitation and benefit changed to become Chile a country with a status quo in this area, for more than 70 years. The final chapter seemed to be closed in 1967, when Borax Consolidated, an English monopoly producing company, which exploited the Surire (Chilcaya) deposits and, subsequently, Ascotán, abandoned its deposits and the country [15]. Subsequently, in 1980 it had a vigorous upturn, and in the coming years, the production of borates and boric acid will be considered strategic because boron is associated with lithium in these Andean saline deposits. This alternative is only a potential alternative because it will depend exclusively on market and non-technological factors.

The Andean sector known as the Code of Santa Cruz (**Figure 2**), localized between 14° and 27° south latitude and 68° 30` west longitude until the Pacific Ocean, has the greatest variety of deposits in relation to that of the neighboring countries. The Andean orogenic system can reach up to 600 km wide and is characterized by numerous closed basins that serve as local base levels for drainage systems in the High Cordillera, also known as Puna or Altiplano [18]. These basins can be of tectonic, volcanic, or other subordinate origin. Their maximum age is probably middle Tertiary, and they continue formed during the Quaternary. Some of them are in full evolution, today with few outcrops of Paleozoic and Mesozoic bases. Its axes, in general, have a meridian to submeridian orientation and, from the Tertiary, have had an active sedimentary development. Some of them can be big, as is the great Bolivian basin, which includes the Poopó and Titicaca lakes, and the great Salar de Uyuni. In Chile, we have the case of the Pampa del Tamarugal [18]. The most outstanding sedimentary aspect of this basin has been the development of lacustrine systems with fresh or saline waters that, in their terminal stages, have evolved into evaporitic basins that are known today as "salars." These are located in the lower part of the basins and, often, asymmetrically within them. The salars

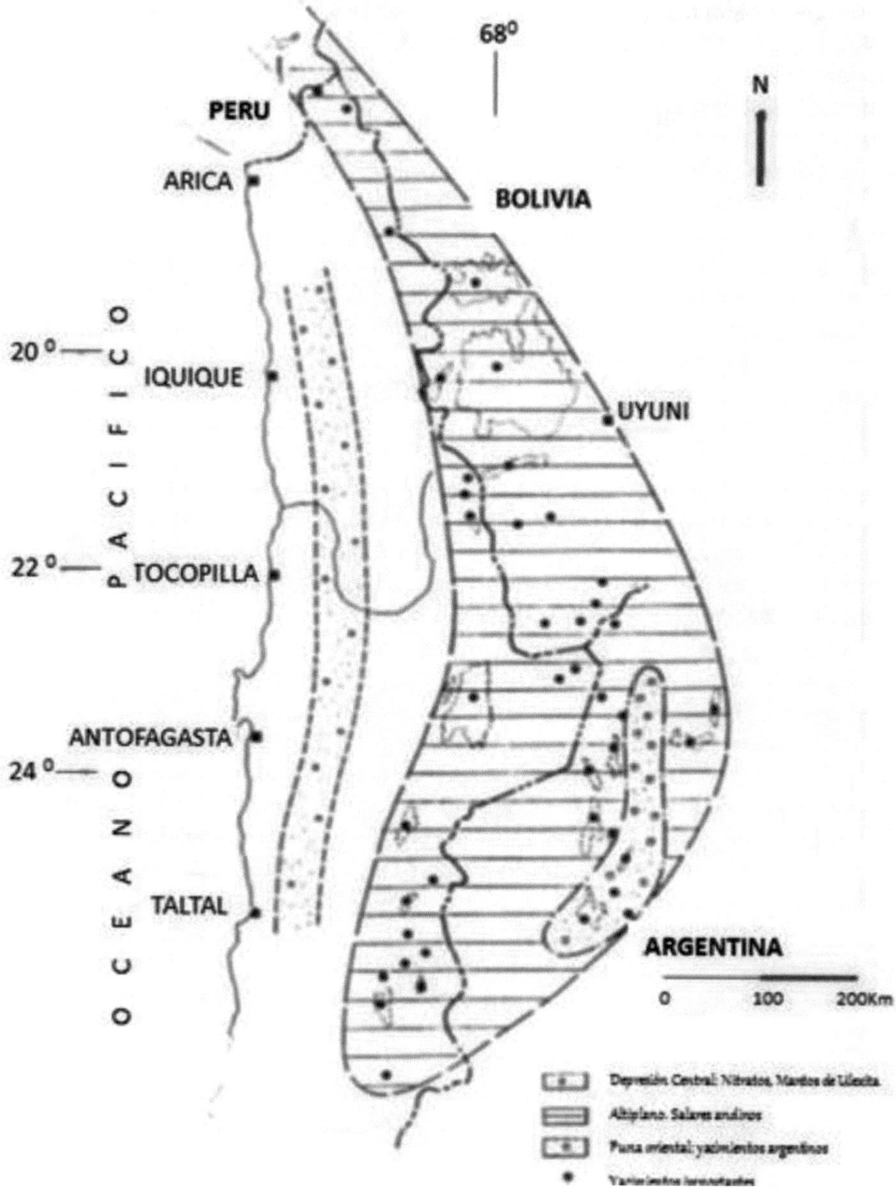

Figure 2.
South American Boratífera Province [9]. The Andean sector is known as the Code of Santa Cruz, that is, approximately between 14° and 27° south latitude. The Andean orogenic system can reach up to 600 km wide and is characterized by numerous closed basins that serve as local base levels to systems of drains of the High Cordillera, also known as Puna or Altiplano. It comprises deposits of Neogenic and Quaternary borates, formed in the last 7 Ma.

result from the breakdown of the balance between the waters contributed to the basins and the evaporation and evapotranspiration processes. A salt flat (salar) is a saline detritic body located in the lower part of closed basins, formed in arid to semiarid environments. In its border, clastic deposits are formed, normally the distal parts of coalescent alluvial systems, which interdigitate with the salts that fill the basins themselves [19]. The contributions are both superficial and underground, with the latter predominant. The salts usually form concentric zones, due to their different solubility product, while the clastic materials are deposited according to their relative granulometry (gravel, sand, clays). Lagoons and groundwater levels are common in salars. The surface of the salt flat (salar) is characterized by salt crusts and structures that reach a great variety, due to the dynamism of the salts. In the major basins, sub-basins with complex hydrological systems have differentiated over time. These basins, although in time and space are very varied, have as a common factor that most of their components, both detritic and chemical, have a majority volcanic origin of Cenozoic age [19]. This includes processes of erosion, weathering, and leaching of volcanic rocks, in addition to direct volcanic activity through the geothermal gradient, pyroclastic materials and lavas that enter the basins, and the contributions of the thermal sources. This is as true for salt flats inside eroded volcanic systems or boilers, such as those formed by tens or hundreds of kilometers of current volcanic systems [20]. The size of the salars varies, from up to thousands of square kilometers, such as the Salar de Uyuni that reaches about 9000 km^2, but its in-depth composition constitutes one of its most unknown aspects. In exceptional cases, such as the Salar de Atacama, it has been recognized that evaporitic bodies can reach more than 1 km of power [20].

The deposits of the high Andean subprovince of the Chilean sector are exclusively of the salar type, and to date, there are no described deposits associated with Miocene sedimentary sequences, as happens in Argentina. The latter are not related to salt flats and present a great mineralogical variety, becoming world-class reserves (subprovince of the eastern Puna) reaching the order of 100 million tons of B_2O_3 [15]. The only deposit described in Chile is the Salar de Maricunga, whose age would be in the "Miocene Superior to the Lower Pliocene" period [21].

The geomorphology of the region corresponds to the block systems in a tectonically active desert. The main blocks, in an integrated profile, from the west, and at a zero level to approximately at 6000 m altitude, are the mountain range of the coast (Cordillera de la Costa); the Central Depression, with its most characteristic section in the Pampa del Tamarugal; the precordillera, with the Cordillera Domeyko Mountain Range; and the Andean Mountain Range (Cordillera de los Andes), formed by the Altiplano and the Cordillera de los Andes. These morphological units may have limits, including sometimes obliterated, or they may be perfectly defined by regional structures [15]. This position geomorphologic to have permitted to classify the salars by your geographic position in Salars of the Coast, Central Depression Salars, Preandean Salars and Andean Salars [22]. The geological framework includes a wide stratigraphic record and igneous rocks, probably the most complete in the country, presenting a great variety of mineral deposits. In this region, the optimal conditions are present for the development of this saline deposits or salars, due to climatic conditions (desert to semidesert environment), geomorphology (abundance of closed basins), and a varied geological framework, both in age and in lithology, in addition to a relative abundance of outcrops [23, 24].

A striking characteristic of this area in southern Peru and northern Chile is the extreme aridity that has prevailed for several million years [25]. Due to this long-lived hyperaridity, the regional landscape experienced extremely low denudation rates (<0.5 m Ma^{-1}), as supported by several cosmogenic nuclide studies [26, 27] in southern Peru and in northern Chile [25, 28, 29], allowing a long-term preservation of geomorphology [30].

Contributions to different basins correspond, mainly, to meteoric and "thermal" waters that, when they reach the basins, of saline composition, evolve into chlorinated brines with lower carbonate and sulfate contents, although the latter can be enriched in situ [31, 32]. The higher cations are sodium, calcium, potassium, magnesium, and lithium, with significant local amounts of arsenic and mercury, while the main anions are sulfates, chlorides, and carbonates, in addition to nitrates and borates [20, 33–35]. Additional factors that contribute in the formation of basins are erosion and leaching of different types of rocks; the transformation, in situ, of rocks by the chemical and physical action of the salts, through "saline tectonics"; biological activity; photochemical reactions; and mists ("camanchacas") and marine spray [36, 37].

It is evident that, from geological times and even in historical times, there is a polarization in aridity, from west to east. As a result, there are "fossilized" salt flats, in the western most part, as is the case of the Salar Grande de Tarapacá (salar of the coast); others receive occasional recharge such as the salars of the Central Depression and Salar de Llamara; others such as the Salar de Atacama (pre-Andean salar) receive contributions from ephemeral rivers and the eastern sector and others that, at present, are in the process of formation like those found in the Andean Mountain Range such as Salar de Loyoques or Salar de Surire (Andean salars).

Another source of boron and lithium are the geothermal fields. In the brines of geothermal fields, there are significant concentrations of salts, including borates. It has been proven that the brines of the Tatio geothermal field have values of the order of 200 ppm of boron, and in similar fields in the United States like Salton Sea, average values of the order of 250 ppm are detected [23, 24]. The basins of the andean salars have a structural control, some have been formed by entrapment flows of lava or inside ancient volcanic systems eroded, for example, boilers. The contributions to the basins are influenced by the volcanic activities that, in some salt flats, are evidenced in its interior by sources of thermal hot springs (Surire, Aguas Calientes I) [19, 35]. The volcanic activity provides a high geothermal gradient that directly affects the quantity and quality of the leached transported ions to the basins [20, 38]. The brines can have up to 220,000 g/l of dissolved solids and, compared to their environment, can reach values 100 times more than those found in meteoric waters and 10 times than those of thermal waters. The main contributions are related to the Cenozoic volcanic rocks, and the brines are mainly chlorides, having as main ions sodium, calcium, lithium, magnesium, chlorides, sulfates, borates, and carbonates [20]. They also have significant amounts of arsenic, mercury and significant strontium anomaly, and, at the level of traces, cesium and rubidium [33, 34, 39]. About 90% of the rocks where boron deposits are formed are volcanic or volcanoclastic, mainly ignimbritas [22].

3. The boron resource in Chilean salars

The different alternatives of exploitation of boron deposits in Chile, either as ore or by-products, in order of importance in terms of magnitude were described by Chong and Garcés [15] and subsequently expanded in Garcés and Chong [22] as:

a. **Deposits in the Andean salars**: The salt flats are located in the high mountain range at altitudes between 4000 and 4500 m.a.s.l. These are distributed in the first three regions of the north. The boron deposits are the biggest reserves in the country, and the predominant geological framework is the Cenozoic volcanism illustrated by the presence of volcanoes and hot springs. The most important starting from the north are Surire, Carcote, Ascotán, Aguas Calientes Norte 2, Quisquiro, Aguas Calientes Sur, Pedernales, and Salar de Maricunga.

Ulexite in Chilean deposits comes in two forms:

- In lenses of powers ranging from centimeters to, in exceptional cases, 2 m. They can extend of a few tens to hundreds of square meters; however, the continuity is quite irregular, equal that its quality or law of mineral. These horizons are called bars.

- In irregularly distributed nodules in the clay sediments known as "potatoes," which do not exceed 15–20 cm in diameter; mostly, these concretions are 10 cm in diameter and have the highest concentrations in B_2O_3 units.

Both types of deposits maintain a genetic relationship with waters from thermal sources. Among the companions of the ulexite are arsenic, gangue minerals such as calcium and sodium sulfates, sodium chloride, and varying amounts of clay minerals or other sulfates and silicates. The presence of other borates such as probertite, its anhydrous equivalent, is frequent, but in subordinate quantities, colemanite and borax have also been detected and don't represent economic importance [40].

The brines in these salt basins also have anomalous borate contents in some cases, which must be discarded from the point of view of eventual economic use. Subsequent works done by Garcés [19, 22, 24, 32], in the Salar de Ascotán, Quisquiro, and Huasco, show that the presence of boron in its brines reaches contents of 550 mg/l, 1627 g/l, and 507 mg/l, respectively.

The only ore mineral in those deposits is the ulexite, and its exploitation is carried out exclusively in the first 2 m of the salt crusts. This limitation is defined by the shallow position of groundwater levels. In the salars of Ascotán and Quisquiro, with the ulexite, probertite, hydroboracite, borax, and colemanite were found. The biggest composition of ulexite in units of B_2O_3 is presented in potatoes, reaching values of 34.74% B_2O_3 compared to 18–19% in B_2O_3, which has the ulexite in bars of the Salar de Ascotán. For Quisquiro samples, potatoes reach 35.43% in B_2O_3, while in bars they range from 9% to 23.8% in B_2O_3. In the Salar de Huasco, a little further north than the previous ones (20°18'S and 68°50'W), probertite, hydroboracite, and ulexite are recognized in a subordinate way; despite the anomalous contents in some sampled points, the salar does not correspond to a boron deposit, since it consists mostly of sulfates, such as thenardite and gypsum, accompanied by halite [19].

b. **Pre-Andean salars:** Those saline deposits are located between the precordillera and the cordillera of Andes. They correspond exclusively to the Salar de Atacama and Punta Negra, both are the oldest, the first of these being the largest and most economically important in the country. Both are part of the Atacama-Punta Negra system. According to Alpers and Whittemore [41], the geological framework is varied with a predominance of Cenozoic, volcanic, and detritic rocks; both salt flats in the past were part of a single tectonically controlled basin. The Salar de Atacama is closed to a great volcanic activity on the eastern side, producing a drainage of the basin with great influence on the geochemistry of its waters, especially of underground origin, through the contribution of CO2 of volcanic magma. The eastern edge of the salar presents horizons of ulexite, associated with gypsum, but in nonsignificant quantities as for its exploitation, usually the presence of the gypsum makes difficult the visual recognition of the ulexite. The brines of the Salar de Atacama are rich in trace elements such as lithium (0.15 g/l) and potassium (1.8 g/l), these being the highest contents that have been found in Chilean salt flats and are currently exploited jointly with boron and sulfates that water brings (see **Table 1**). These brines are concentrated until boric acid is obtained through a process of concentration by solar evaporation. The boric acid produced is a product of the recovery process of potassium sulfate that brine is extracted in boron concentration. Subsequently, the brine is acidified with sulfuric acid to obtain the crystallized boric acid. The crystals are filtered

Species	Boron	Chloride	Sulphate	Sodium	Potassium	Lithium
S. Atacama© % p/p	0.06	16	1.78	7.6	1.8	0.15
S. Surire (mg/l)	1627	144,028	13,583	82,225	13,500	472
S. Punta Negra average value (mg/l)	0.0377	54.125	4.17	34.2	0.845	Not analyzed
S. Aguas Calientes N (mg/l)	729	117,655	1984	71,581	1831	227
S. Maricunga average value (mg/l)	596	190,930	709	85.190	8237	1123

© Jacobs Engineering Inc. for Minsal Ltda, 1989.
Source: [19, 42, 48].

Table 1.
Chemical composition in mg per liter of the brines in Andean Salar's, except Atacama salar's.

and washed in counter current in three successive stages, to finally dry them and obtain the boric acid of high purity. For other parts, the Salar de Punta Negra is the second biggest in the country, located at an altitude of 2956 m, close to 200 km south of the Salar de Atacama, between the Cordillera de los Andes and the Cordillera de Domeyko. The salar drainage basin is approximately 4660 km^2. The closest volcanoes to the east edge are the Llullaillaco (6723 m) and the Socompa volcanoes (6031 m). The geological framework consists of lithostratigraphic units, which include almost all of the statigraphic columns, since marine and continental, volcanic, metamorphic, and intrusive sedimentary rocks are present, ranging from Paleozoic to Recent. In some sectors in its eastern part, ulexite lenses, intercalated with sediments and crusts of calcium sulfate, show small-scale exploitation in the past, but today no eventual future of exploitation of this resource is estimated. Studies of Gannat and Schlund [42] indicate the western edge of the Salar de Punta Negra is mostly composed of halite and gypsum and in a subordinate way at the level of traces, glaserite, nitratite, potassium nitrate, humberstonite, silvite, darapskite, tenardite, glauberite, singenite, and polyhalite.

c. **Deposit saline of the Central Depression:** The Central Depression is a depressed relief that extends between the Cordillera de la Costa and the Precordillera. The salt flats found in this plateau are closed watersheds of the endorheic type, which receive water contributions from the east (Alta Cordillera), underground and surface. The salt concentration mechanism is based on the leaching of these materials, associated with mud flows or other irregular water avenues, which are transported to the Central Depression and are the formators of this type of salt flats in the distal part of alluvial cones or isolated bodies in some basins [22]. These deposits include, at the base, decimetric sequences of clays and saline silts and monomineral decimetric horizons of tenardite, humberstonite, bloedite, ulexite, and darapskite. It should be considered that the parts of these mud flows have volcanic origin. Characteristic deposits of this type are found in Cerros de la Joya and Cerro Antar in the southern part of the I Region of Tarapacá (21° 50′ 32″ S and 69° 27′ 17″ W), which were exploited with some intensity in the past. In the same area, in the upper part of some salt flats, lenticular bodies of centimeters thickness of ulexite are recognized. These have also been exploited sporadically, in the area of the Salar de Pintados in the first region was also exploited in the past, but in general, in all these deposits, they can consider the ulexite of subordinate economic importance. In the Pampa del Tamarugal between coordinates 19° 40′ S and 69° 40′ W, the minerals found in boron correspond to ulexite, associated with halite, potassium salts, and carbonates, in beach-type deposits.

d. **Deposits of borates and by-products from the salt industry:** This economic alternative to recover borates from the salt industry requires the exploitation of large volumes of nitrate ores, and the recovery of borates would be a by-product of this industry. The geology and mineralogy of these nitrate deposits are described in various authors [31, 43, 44]. George Ericksen studied these deposits in detail, from a structural, geological point of view, their history, mineralogy, and geochemistry. He found and described several iodine minerals he called brüggenite, hectorflorecite, fuenzalidite, and carlosruizite, among others. Ericksen [44] includes the presence of borax in the nitrate deposits, together with inyoita, kaliborita, and gowerita but in low relations. Inderite, ginorite, hydrochloroboracite, and hydroboracite have also been recognized, but all of them in minimal quantities. Minerals recognized in the field of nitrates are halite, silvite, sodium nitrate, potassium nitrate, darapskite, humberstonite, tenardite, anhydrite, bassanite, gypsum, kieserite, epsomite, glauberite, bloedite, apthitalite, picromerite, lautarite, ulexite, and probertite hydroboracite.

The minerals treated from caliche in the Pedro de Valdivia and María Elena plants contain contents in B_2O_3 units over 1%, of the minerals treated in plants, during the period from 1932 to 1967. Data are currently referenced in Wisniak and Garcés [45].

e. **Brine by-products in geothermal fields:** It is known that the waters of geothermal fields, such as the case of Tatio, for example, are enriched with borates (183 mg/l) [46] as in other salts of economic importance (even of metallic elements). The economic recovery of borates must be considered in this case as secondary in all its aspects, since it depends on the initial exploitation of geothermal energy and the salts contained in the brines. Currently there is no project that estimates in the near term geothermal exploitation and would not involve large volumes of production.

4. Description of the area of the deposits studied

Chile only occupies a narrow strip, which extends from 18° to 27° south latitude and 68° 30′ west longitude, from the South AmericanBoratífera Province (**Figure 2**); but these appear in all the latitude of the northern part of the country, from the eastern limit to the coast. The westernmost part is a desert domain (Atacama Desert in the strict sense), while, to the east, it can be considered semi-desert (Puna or Altiplano). The aridity of this area, however, is very special, due to the winter rainfall of the Altiplano, added to those of the summer known as "Bolivian winter," cyclically carrying large amounts of water to the east sector. This means a considerable recharge, in addition to mud currents and ephemeral lakes that can reach the coastal edge. All these waters drain into the interior drainage basins that have intermediate base levels. The geomorphology of the region corresponds to block systems in a tectonically active desert. The blocks, corresponding to main reliefs, are controlled by faults of meridian to submeridian heading, and another transversal EW. The main blocks, in an integrated profile, range from west to zero level up to 6000 m altitude.

Salar de Surire: The salar is located northeast of Chile at 18° 53′ S and 69° 03′ W at an altitude of 4300 m with a 140 km^2 closed basin. The most characteristic is in the central part because the volcano Oquecollo emerges to a height of 70 m on the surface cutting the saline cover [19]. The salar receives water from two forms, meteoric and groundwater, creating a very shallow body of brine in the south central area. The region is semiarid with an average annual rainfall of about 230 mm [19]. During the rainy season, the salar may be brine covered to a depth of 20 cm, but during the dry season, the brine water table in the dry season cements the upper crust pores. There are a number of saline thermal springs located on all southeast border of salar, the principal is Polloquere. Almost all the time the river is dry up during the dry season. This suggests that it is fed mainly by atmospheric precipitation. Each year rainwater partially recharges the groundwater system which in turn discharges in topographic depressions. Such waters don't undergo deep circulation, which would put them in contact with ancient evaporites. They owe their composition almost exclusive to the weathering of the volcanic rocks and the resolution of ancient buried evaporites.

The mean annual air temperature is estimated between 10 and 15°C. Air temperature ranges from −15°C in winter to 20°C in summer. Daily variations may reach 35°C. The potential evaporation is high 2500 mm/year [36]. The deposits that make up the salar correspond to carbonates, sulfates, chlorides, borates, and sinter deposits [19]. The Borax Chemical and Industrial Company operates the ulexite deposit in the Salar de Surire since 1985, producing a range of boron products, including boric acid, boron granules (known as granulex), and fertilizers and insecticides, which combine boron or boric acid with calcium sulfate and sodium borates [47]. The surface morphology of the salar is practically flat, although with a slight tilting toward the west, conditioning that the waters that access it go and accumulate in its western margin. Yellow and red colorations are observed that are assigned to arsenic sulfides [19] which reveals a markedly arid environment.

A relevant aspect of the hydrology of the Salar de Surire is the existence of a swampy halo rich in vegetation, which is in the marginal area of the saline body that constitutes the salt flat. It is a special ecosystem known as bofedal and that consists of a high-altitude wetland where varied and abundant vegetation develops. This ecosystem coexists in the middle of a large area of salts, mainly borates that affect the development of its vegetation and others such as arsenic that give it a certain toxic character. The importance of these bofedales is the water they contain, and that is essential for the varied biodiversity to exist into Andean altiplane.

Salar de Quisquiro o Loyoques: It is located in the Altiplano of the Antofagasta region between the latitudes of 23° 07′ and 23° 30′ south latitude and the longitudes 67° 10′ and 67° 30′ west, at 4430 m altitude. This saline deposit is a closed basin of about 80 km^2; it is a beach-type salt flat with few surface lagoons of variable extension (around 5 km^2) and a brine of a few decimeters deep. The main characteristic is that it presents in its center a borate mine with exploitable levels of ulexite, which in the past was exploited [22, 23]. The main surface contributions in this salar are the Salado River in the south and in the north the Loyoques estuary, in addition to the three-deep pass in the west (from north to south) Agua Escondida, Taina, and Quisquiro. The runoff (tributary to the salar) is 3500 l/s; the surface evaporation of the waters is 1500 mm/year, and the average annual rainfall is 150 mm [36]. These data show that the high evaporation rate is due to the high aridity in the area. The maximum average temperature ranges between 10 and 20° C, while the minimum average temperature is a few degrees below 0° [36]. The winds in the altiplane follow a pattern, which in the morning is of low intensity with north orientation, and in the afternoon they start from moderate to high intensity, with west to east orientation.

This saline deposit shows an overload exceeding a continuous horizon of ulexite that varies between 1 and 2 m of power with halite and small amounts of arsenic sulfide and native sulfur particles. In its upper part (30–40 cm), despite the homogeneous aspect of the mineralization, it presents a low grade not exceeding 12% in B_2O_3 in dry sample. At the same time, in its lower part, eliminating interstitial water can reach concentrations of up to 24% in B_2O_3 [4]. At the edges of the salt flat, the overload is dry and hardened presenting gypsum crystals and halite with clay and sand. The underlying ulexite, mixed with abundant gypsum, surpasses sterile clay horizons. The sections that do not contain mineralization are formed by hard salt crusts (20–30 cm) that overlap clays and sands with abundant organic material under which it is found, the water table [22, 23].

The hydrological characteristics are controlled by the conditions of zonal aridity and affected by summer rainfall from the Amazon Basin. This endorheic type basin is characterized by presenting potential resources for water use from its confined layers, as well as mineral species of commercial interest (B, Li, etc.).

Salar de Aguas Calientes I: Evaporitic deposits are located at coordinates 23° 07′ 22° S and 67° 26′ W, at an altitude of 4280 m and have an extension of approximately 15 km^2. It drains to an endorheic basin of about 281 km^2, with an orientation of its main axis NS of 6.5 km and that in its widest part reaches 2.4 km. It is a beach-type salt flat, with surface lagoons of variable extension which occupies a relatively small area of depression, and is covered by a crust of saline materials, mostly halite and gypsum [36, 48]. This situation is due to the current water balance, clearly deficient, since the rainfall received by the drainage basin is around 150 mm/year, with the estimated evaporation exceeding 1500 mm/year which reveals a clearly arid environment. The surface morphology of the salt flat is practically flat, although a slight tilting toward the west determines that the waters that access the salt run and accumulate in its western margin. The geological substrate of the area where the salt flat is located consists mainly of igneous rocks of the Pliocene age. The most important is the ignimbrite atana with a volume of about 2500 km^3 of white, pink, and gray daffitic tuffs [48]. It belongs to the geological zone known as the Altiplanic Puna Volcanic Complex (APVC). This unit is stratigraphically located on other older ignimbrite units, associated with the igneous activity existing in this sector at least from the Upper Miocene [49, 50], associated with the La Pacana caldera, which has been interpreted as the oldest collapse boiler of the Cenozoic volcanism in this region. The combination of climatic aridity and high altitude translates into rigorous environmental conditions, among which the following stand out: low partial pressure of oxygen and carbon dioxide, high solar radiation, poorly developed soils with low nutrient availability, low temperatures with a sharp daily oscillation, irregular distribution of rainfall, and prolonged periods of aridity. Under these conditions, the salt flat and its surroundings constitute a crucial water reserve for the development of life.

From an economic point of view, the salar has important reserves of borates, highly soluble mineral salts, which are genetically related to the volcanic activity of the Tertiary Superior-Quaternary [10, 50]. The boron deposit in this salt flat has been prospected, but not exploited, leaving on the surface of its central zone a series of open calicatas of up to 2 m deep, at whose bottom the groundwater of the salar.

5. Geochemical modeling

The physical–chemical processes, including the solubility or deposition of various salts, are known to be determined by the ionic composition of the solutions. The solubilities of minerals in brines can be calculated from thermodynamic considerations, provided the equilibrium constants are known and activity coefficients can be obtained. In this context, we have used Pitzer's ionic interaction model with the extension formulations of the Harvie and Weare model [51–53] to predict the precipitation of salts in multicomponent aqueous systems with high ionic strength in a temperature range of 0–60°C [54–56], using three salt natural brines of Andean borates.

The code used for the thermodynamic calculations of the brines studied in this work is the PHRQPITZ [57], which incorporates and extends to temperature ranges from 0 to 60°C with the Pitzer approximation and the parameterization performed by Harvie et al. [53] for the Na-K-Mg-Ca-H-Cl-SO_4-OH-HCO_3-CO_3-CO_2-H_2O system. The codes are subject to compliance with conditions of chemical equilibrium and conservation of mass and/or ionic. These conditions translate into a series of government equations in which knowledge of the activities of the species present is essential. This software calculates the individual ionic activities from the analytical concentration data, based on the specific interaction method. Subsequently, it evaluates the degree of saturation of the solutions with respect to the mineral phases of interest and compares the products of ionic activity (PAI) obtained against the equilibrium constants (Keq) corresponding to the minerals, by means of the saturation index (IS) through the expression:

$$\text{I.S.} = \log\ (\text{P.A.I.}/\text{Keq}) \tag{1}$$

where P.A.I. represents the product of ionic activity and Keq the corresponding equilibrium constant. If the I.S. shows zero value means that the solution is in equilibrium with respect to that phase, on the other hand, positive or negative values of the I.S., they determine a situation of supersaturation or subsaturation, respectively, whose magnitude is a direct function of the absolute value of the quantity.

The solubility of a mineral in equilibrium is represented as:

$$X^aY^b \bullet nH_2O \Leftrightarrow aX^{b+} + bY^{a-} + nH_2O \tag{2}$$

If the solubility product of a mineral is defined as (Ksp, X^aY^b•nH_2O) and it is considered as the solid phase activity equal to 1, then.

$$\text{Ksp, XaYbnH}_2\text{O} \Leftrightarrow a^aX \bullet a^bY \bullet a^nH_2O \Leftrightarrow (m_{x,sat} \bullet \gamma_X)^a \bullet (m_{y,sat} \bullet \gamma_Y)^b \bullet a^nH_2O \tag{3}$$

where $m_{x,sat}$ and $m_{y,sat}$ are the cation and anion mole concentrations, in the saturated liquid phase with respect to the solid phase. For example, for gypsum, Eq. (2) looks like:

$$CaSO_4 \bullet 2H_2O \Leftrightarrow Ca^{2+} + SO_4{}^{2-} + 2H_2O \tag{4}$$

The gypsum solubility product (Kps, $CaSO_4$•$2H_2O$) depending on the activity is:

$$\text{Kps} = (a_{Ca^{2+}} \bullet a_{SO_4{}^{2-}} \bullet a_{2H_2O})/a_{CaSO_4 \bullet 2H_2O} \tag{5}$$

The activity of the solid (mineral) phase is defined as equal to 1 ($a_{CaSO_4 \bullet 2H_2O} = 1$), where Eq. (5) is transformed to:

$$\text{Kps} = (m_{Ca^{2+}} \bullet \gamma_{Ca^{2+}}) \bullet (m_{SO_4{}^{2-}} \bullet \gamma_{SO_4{}^{2-}}) \bullet a^2H_2O \tag{6}$$

This equation can be rewritten as:

$$Kps = Kps^{*} \bullet \gamma_{Ca^{2+}} \bullet \gamma_{SO_4{}^{2-}} \quad (7)$$

where Kps* is the stoichiometric solubility product, $Kps^{*} = (m_{Ca^{2+}} \bullet m_{SO_4{}^{2-}})$. In the equilibrium, the Kps depends on the thermodynamic and stoichiometric product. The activity coefficient (γ) and the solubility depend on several factors, the main one being the ionic force (F.I.). However, as the activity coefficients decrease with the ionic strength of the medium, the Kps value should increase as a function of the ionic strength that the thermodynamic solubility constant does not vary. This increase in stoichiometric constant (Kps) translates into an increase in solubility, as can be deduced from Eq. (7); this is known as the "saline effect" [58].

6. Material and methods

Sampling of surface was collected in the Salar de Surire, Quisquiro, and Aguas Calientes I. Samples of water are taken for analysis from river, springs, and pool, during two seasons, summer and end of winter, at the same previously labeled extraction points and geographically located with GPS. During transport to the laboratory, the samples were kept refrigerated, protected from light, and sealed. The pH, temperature, TdS, and Eh were measured in the field with portable devices pH-conductivity meter (Orion Star A325) which is periodically calibrated with buffer solutions. After making these measurements, the samples were filtered through 0.45 μm filter papers and preserved for analysis. Each sample was stored in two polyethylene bottles. One of them was acidified with 10 mL HCl for cation analysis. The other was kept unacidified for Cl, SO_4, and HCO_3 analyses. Samples were stored at 4°C for being envoy to the laboratory analyses. In the laboratory the cations (K, Na, Ca, Mg, Li, and As) were determined by atomic absorption spectrophotometry (Perkin Elmer 2380), using the standard protocol, chlorides by the Mohr method, and SO_4 by gravimetry with BaCl and volumetric analysis (Cl, B_2O_3, CO_3, HCO_3). Sulfate determination was carried out by precipitation with $BaCl_2$, using the Mohr method for the analysis of chlorides [59] and the acid-base technique for borates [60].

The quality of the analytical data is evaluated by calculating the percentage of charge-balance error, by the expression of Ball and Nordstrom [61]. In parallel, samples of saline scabs and sediments are taken, in which, once dried, their mineralogy is analyzed with a atomic absorption diffractometer (Siemens, mod. D5000):

$$\%\,Error = \frac{\sum (m_i z_i)_{cationes} - \sum (m_i z_i)_{aniones}}{(\sum (m_i z_i)_{cationes} + \sum (m_i z_i)_{aniones})1/2} * 100 \quad (8)$$

With the data of the chemical analysis of the brines, the geochemical modeling calculations are carried out, using a speciation code. The values of ionic strength (F.I.), water activity, ionic activities, and saturation indices (I.S.) of the solutions with respect to the salt phases of interest are determined. The geochemical modeling code PHRQPITZ [57] was used, the most suitable for the treatment of concentrated solutions. The values of the equilibrium constant of thenardite (Na_2SO_4) are included, because it was not in the original database of the software, being of great interest in natural brines with high concentrations. Similarly, the values, at different temperatures, of the equilibrium constants of other mineral phases are of interest such as glauberite ($Na_2Ca(SO_4)_2$), a mineral that appeared in the original database of the PHRQPITZ code only at 25°C.

The accuracy achieved in the calculation of saturation indices depends on the quality of both the thermodynamic parameters handled (included in the code database and those from the necessary extensions, such as those commented on the equilibrium constants) and the analytical results obtained. The criterion applied to assess uncertainty has been ±0.15 units of I.S., implying that equilibrium situations cannot be determined with greater precision than defined [62]. In the case of carbonated minerals, a slightly higher range of uncertainty is considered, besides ±0.4 units of I.S. This different treatment is due to the existence of important methodological problems related to the determination of pH in highly concentrated solutions [63, 64], because the use of conventional buffers for the calibration of the pH meter determines the measurement of this parameter in brines is not in the same scale of activity coefficients as the aqueous model used [57]. This type of problems related to the behavior of the carbonated system in high-concentration solutions has been manifested even in experimental studies [65].

The evaluation of the saturation status of the solutions against the selected mineral phases is performed by calculating the I.S. The equilibrium constants used are those that appear in the database of the code used. Examples of similar studies are [35, 66–68].

7. Results and discussion

The analysis of the data from the water samples collected in the different salt flats allows us to express certain considerations that are detailed below.

The representation of the chemical composition of the surface water samples is represented in **Figure 3**. It shows the similarity between the water contributed from the thermal springs and the surface brines of the salt, because the thermal waters arise from the interior of the salt flat and, when dissolving salts in their ascent, acquire ionic proportions similar to the surface brines sampled inside the salar. On the other hand, the diluted waters that provide the slopes show greater ionic variability, because the salt drainage basins are made up of different types of volcanic and sedimentary rocks, which give a different chemical signature to the waters that flow about them. The waters of the Salar de Surire present greater variability of chemical types (with a greater number of samples taken) as well as the Salar de Quisquiro, while the Salar de Aguas Calientes I presents the most homogeneous waters.

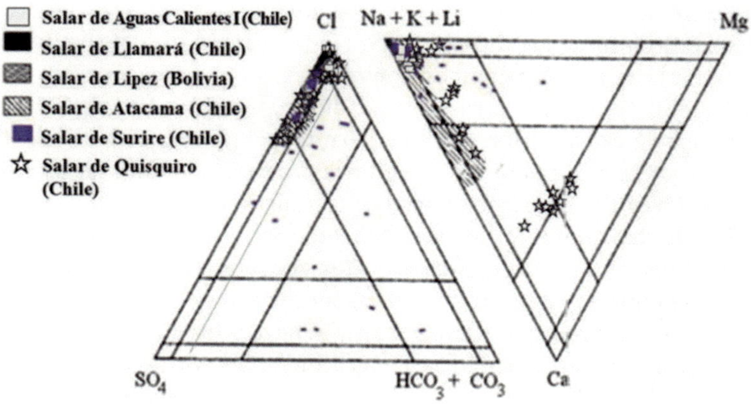

Figure 3.
Graphic comparison of the chemistry of the waters of the Salars of Surire, Quisquiro, and Aguas Calientes I with the Salar de Atacama and Bolivian salt flats.

In the Salar de Surire, the brines accumulated in its internal surface exhibit a lower variability, the anionic composition of the Cl-(SO_4) and Cl type, and the very homogeneous cationic of the Na type. On the other hand, the samples of springs present a high degree of variation in the distribution of both cations and anions. The anionic composition is of the HCO_3-(SO_4)-(Cl) and Cl-(SO_4) type, with virtually all intermediate types, and the cationic varies from the Mg-Na-(Ca) type to the Na-Mg type. These waters are slightly alkaline, with a pH ranging between 6.2 and 9.7. On the other hand, the waters of the Salar de Quisquiro are of the Na-Cl-SO_4 type. Its anionic relationship is Cl- > SO_4^{2-} > HCO_3^- > BO_3^{3-} > NO_3^{3-} > F^-, and its cationic relationship is Na^+ > Ca^{2+} > Mg^{2+} > K^+ > As^+ > Li^+. The description of ionic compositions and their temporary changes in closed natural water bodies allows for an explanation of the formation mechanism and the development of enrichment of its brines. The flow of water that enters in this closed basin shows a considerable variety of ionic composition, like in this concentration, the pH varies between 7.2 and 9.02. Finally, the chemical representation of the waters of the Salar de Aguas Calientes I tells us that these are quite homogeneous; their anionic relationship is Cl- > SO_4^{2-} > BO_3^{3-} > HCO_3^-, and their cationic concentration is Na^+ > Ca^{2+} > K^+ > Mg^{2+} > Li^+ > As^{3+}. They are practically neutral waters, whose pH values range from 6.4 to 7.9. The chemical nature of these waters (**Figure 3**) shows that the salar belongs to the Na-Cl system, and if we consider the equimolar relationship between [Na] and [Cl], they are practically the same, which has a direct relationship with the saline composition of the salar; this suggests that the high concentrations of Na and CI of the most saline waters come essentially from the solution of sodium chloride or halite (NaCl). Therefore, in the most saline contribution waters, the excess of Na and Cl (in relation to the more dilute waters) probably comes from the redisolution of old evaporites associated with sedimentary rocks covered by the extensive volcanic formations of the east and northeast of the basin. When comparing the chemical nature of the waters of the Salar de Aguas Calientes I with the Lipez Bolivian salars [39] on the diagram (**Figure 3**), it can be verified that the brines of the Lipez salars are of the type Na-(Ca)-Cl-(SO_4) and vary to Na-Cl type while the waters of the Salar de Aguas Calientes I, show little variability of the type Na-Cl-(Ca), which leads to suggest that their brines contain more calcium than sulfate. In short, the similarity and contrast of the chemistry of our study [48] with that carried out by Risacher et al. [36] 20 years later are complementary.

The thermal springs that emerge inside the Salar de Surire have an anionic composition between the types Cl-(SO_4), Cl-(HCO_3), and Cl, while its cationic composition varies between the types Na-(Ca) and Na. The values measured in the thermal springs in Surire range between 84°C, pH 6.82, and F.I. 0.0832 m and 32°C, pH of 7.84, and F.I. 0.0897 m, while the thermal springs located south of the Salar de Aguas Calientes (black lagoon) has a composition of the Cl-(SO_4) and Na type, with a concentration of 0.46 molal. It is the most acidic of all waters with a pH 6.4 to 47.3°C. From these data we can mention that the thermal springs that occur in the Salar of Surire and Aguas Calientes I are waters of low concentrations and rather neutral.

The evolution of the physicochemical conditions was evaluated with the PHRQPITZ code that allows to calculate the ionic activity coefficients (γ_i), activities of the species present, and the degree of water saturation (I.S.). **Figure 4** shows the evolution of the ionic activity coefficients as a function of the ionic strength. From these graphs it is observed that there is a first group, whose activity coefficients decrease to a minimum and then increase as it increases with the F.I. in the solution; they are γNa^+, γMg^{2+}, γCa^{2+}, and Cl^-. The second group is the γK^+, γSO_4^{2-}, and γHCO_3^-, which is characterized by a progressive behavior with the decrease in F.I.

The saline waters of the Salar de Surire are between 0.004 m to 5.03 m, with values in the calcite saturation index practically supersaturated in this mineral.

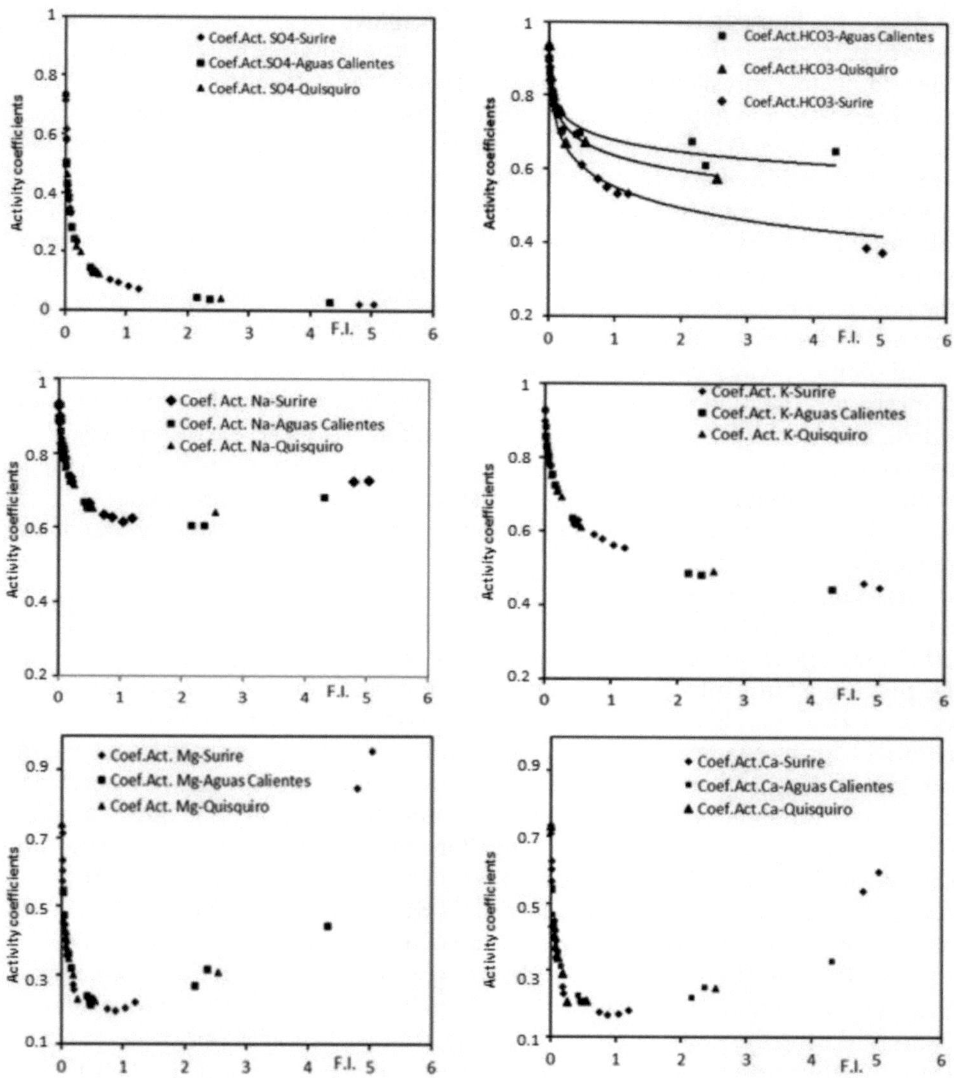

Figure 4.
Activity coefficients of major ions based on ionic strength, calculated from the speciation code PHRQPITZ.

Water samples with F.I. < 0.07 m is far from equilibrium. If we consider alkalinity, such as (alk) > 2(Ca^{2+}) in the initial water entering the salar, then the alkalinity should increase regularly during the evaporation process and only precipitate calcite, so the waters will follow the alkaline pathway (increasing the carbonate concentration and decreasing calcium). Our waters have a slightly alkaline pH, so the process goes in the direction of the precipitation of magnesium silicates, releasing H^+ ions, neutralizing a large part of the carbonate in the solution, and modifying the Ca/CO_3 ratio, therefore changing the alkaline pathway to the neutral saline pathway, while other waters (to a lesser extent) go through the alkaline pathway. It can be said that if the alkalinity of the initial solution is greater than the total concentration of the calcium (in milli-equivalents per liter: meq/l) [36], the solution will begin its evolution toward the alkaline route, raising the pH above 10. But, if the alkalinity is less than the sum of the calcium and magnesium concentrations (in meq/l), the solution will change its evolution toward the neutral saline route due to the precipitation of magnesium silicates, pH less than 9. Interestingly, it is possible to determine immediately, from the analysis of the initial solution, which evolutionary path should theoretically follow the solution from the precipitation of calcite and magnesium silicates. For its part, as an explanation of the presence of calcium sulfate and gypsum, since there is a clear tendency to balance in the samples that have F.I. greater than 2 m, in this case the existence of the sulfate source is the additional sulfur brought to the waters of the aquifer where it is oxidized transforming into sulfate from the sulfated minerals in the area. But, if the alkalinity is less than the sum of the calcium and magnesium concentrations (in meq/l), the solution will change its evolution toward the neutral saline route due to the precipitation of magnesium silicates, pH less than 9. The interesting thing about this method is that it allows you to immediately determine, from the analysis of the initial solution, which evolutionary pathway should theoretically follow the solution from the precipitation of calcite and magnesium silicates. If all alkali cations such as sulfate are removed, the amount of oxidized sulfate would be the difference between the total loss of alkali cations and the apparent loss of sulfate. From the visual observation in the area, there is a very close native sulfur, which is transported by the winds; probably this is the process that generates the significant reduction of alkalinity.

The Salar de Quisquiro presents concentrations between F.I. 0.004 m and 2.54 m, and all sampled waters saturated in calcite and aragonite are presented. The other minerals closest to equilibrium are gypsum and later magnesite but far from precipitating. Since all waters are supersaturated in calcite, the next step is to consider alkalinity. If (alc) > 2(Ca^{2+}), in the initial inlet water, then it increases regularly during evaporation, if only calcite precipitates. In other words, the activity product of the Ca^{2+} ions in solution in relation to CO_3^{2-} must remain constant, as the concentration increases. If one of the solutes increases, then the other should decrease. In our case, the distribution of all incoming water must be converted into carbonate-rich brines. However, the evolution of the water represented in the triangular diagram (**Figure 3**) is of the Na-Cl-SO_4 type with a pH lower than 9. The usual explanation for this anomaly is the precipitation of Mg silicates [69]. This suggests that the distribution of the incoming waters should become alkaline but that they are neutral brines. This induces evaporitic waters to follow two evolution paths. If during the evaporative process they are enriched with carbonates, becoming an alkaline brine of sodium carbonate, they will have a pH greater than 10. The other way is when the waters of input present more calcium than carbonate so that the solution is concentrated in calcium producing neutral brines with pH less than 9, as it is in our case. Therefore, the fundamental relationship if (alc) > 2((Ca^{2+}) + 2 (Mg^{2+})) does not depend on the nature of the weak anion of the acid associated with Ca and Mg, during the evaporation concentration process. Another explanation of the data is the existence of an excess sulfate of origin. In this model, additional sulfur is put into the waters of the aquifer where it oxidizes to sulfate and removes sulfate minerals. If every alkali cation is removed as sulfate from the amount of oxidized sulfate would be the difference between the total loss of alkali cations and apparent loss of sulfate. In observing the area very close to it, there is native sulfur. This is transported by the winds. These processes give a significant reduction in alkalinity.

The Salar de Aguas Calientes I has concentration waters that vary between 0.0237 m and 4.3195 m. The chemical nature of the waters shows (**Figure 3**) that its waters are of the Na-Cl type and it is directly related to the salt composition of the salt, whose sediments are mainly halite accompanied by gypsum and with the presence of borates in its central zone. The composition of all its waters seems to be linked to the saline composition of the sediments of the salt flat, since its contact with both the surface waters that access the salars and the underground contributions cause the enrichment in chloride and sodium, thus generating a more homogeneous composition of the solutions. The only samples that differ from this behavior are those of the northern sector, presenting a slightly higher anionic proportion in HCO_3 and SO_4. According to the results of the samples, the highest concentration of As and B in the brines is in the central-western sector of the salt, which is due to the predominantly hydrothermal origin of the recharge solutions. The lagoons of the central western sector are waters with a variable degree of chemical concentration. These lagoons are located mainly in the western zone of the salar, due to a slight slope of its surface toward the sector W, accumulating in this margin the scarce rainfall and also the solutions coming from the subsurface flow of the salar like that of its internal zone.

The predominance of water from inside or outside the saline nucleus of the salar determines the greater or lesser degree of concentration of the solutions, although they all share the same chemical type. In the southernmost zone, thermal waters emerge with a degree of concentration in solution similar to that of the brines that accumulate in the nearby lagoons, from which it follows that the concentration of the solutions comes in this area controlled by the interaction with the saline sediments of the salar. The saturation conditions (I.S.) of the solutions analyzed with respect to saline minerals of interest show a variable behavior; in the northern sector, the waters are clearly subsaturated with respect to gypsum and halite, while the waters of the bofedal are saturated with respect to calcite. On the other hand, the hot springs are subsaturated with respect to these three main saline minerals, although the degree of subsaturation is lower than that of the nonthermal springs. Finally, of the samples collected in the lagoons of the central sector, the two most diluted are in equilibrium with calcite and subsaturated with respect to gypsum and halite, while the most concentrated appears to be supersaturated with respect to calcite and gypsum and subsaturated with respect to halite.

8. Conclusion

Chilean salt flats have only ulexite as an economic mineral, these deposits being related to a shallow saline lake, whose main contributions in boron are from thermal sources. Boron in Andean salars is mainly linked to the Cenozoic volcanic rocks that prevail in the eastern Andean territory, particularly in recent or latent volcanic centers, and there is an increase in boron in areas of active latent volcanism, such as the hot springs of Salar de Surire [47] and Salar de Aguas Calientes I [48].

Author details

Ingrid Garcés Millas
Department of Chemical Engineering and Minerals Process, Faculty of Engineering, University of Antofagasta, Antofagasta, Chile

*Address all correspondence to: ingrid.garces@uantof.cl

IntechOpen

© 2020 The Author(s). Licensee IntechOpen. This chapter is distributed under the terms of the Creative Commons Attribution License (http://creativecommons.org/licenses/by/3.0), which permits unrestricted use, distribution, and reproduction in any medium, provided the original work is properly cited.

References

[1] Kistler RB, Helvaci C. Boron and borates. In: Carr DD, editor. Industrial Minerals and Rocks. 6ª ed. Sociedad de Minería, Metalurgia y Exploración, Inc.; 1994. pp. 171-186

[2] Garrett DE. Borates. In: Handbook of Deposits, Processing, Properties and Use. London: Academic Press; 1998. p. 483

[3] Helvacı C. Borates. In: Selley RC, Cocks LRM, Plimer IR, editors. Encyclopedia of Geology. Elsevier; 2005. pp. 510-522

[4] Garcés I, Reyes J, Geraldo C, Yunis F. Estudio experimental de la solubilidad de ulexita en agua y determinación de correlaciones generalizadas para medir su densidad, viscosidad e índice de refracción. Revista Innovación. 2003;**15**: 35-43. ISSN: 0716-6311

[5] USGS. Boron. In: Survey USG, editor. Mineral Commodity Summaries. 2019. pp. 36-37. Available at: https://prd-wret.s3-us-west-2.amazonaws.com/assets/palladium/production/s3fs-public/atoms/files/mcs-2019-boron.pdf

[6] Helvaci C, Palmer M. Origin and distribution of evaporate borates—The primary economic sources of boron. 2017. Available at: https://eprints.soton.ac.uk/413247/1/Helvaci_revised_April_24.pdf

[7] Alonso R, de los Hoyos L. Boratos. Capitulo VII: otros minerales (Coord.) Liliana N. Castro y Ricardo Melgar. In: Nielson H, Sarudiansky R, editors. Minerales en la Agricultura en Latinoamérica. 1a. ed. Buenos Aires, Argentina. 2005. pp. 534-545, 578. ISBN: 987-22647-0-8

[8] Alonso R. Los yacimientos mundiales de boratos neógenos y claves para su prospección en los Andes Centrales. UNAS-Conicet. Conferencia en el marco de su incorporación como Académico Correspondiente Argentino a la Academia Nacional de Ciencias de Buenos Aires, Argentina; 2017

[9] Alonso R. Ocurrencia, posición estratigráfica y génesis de los depósitos de boratos la Puna Argentina [tesis doctoral]. Salta, Argentina: Universidad Nacional de Salta; 1986. p. 196

[10] Alonso R. Los Boratos de la Puna. Edición Cámara de la Minería de Salta. Salta, Argentina. 1998. p. 196

[11] Alonso R, Viramonte J. La cuestión genética de los boratos de la Puna. In: Proceedings of the XII Congreso Geológico Argentino (Mendoza), Tomo V, Buenos Aires. 1993. pp. 187-194

[12] USGS. Mineral Commodities Summary. Boron. 2013. Available at: https://minerals.usgs.gov/minerals/pubs/MineralCommoditiesSummary/Boron/2013

[13] Parra R, Morales M. Estudio básico para la obtención de ácido bórico a partir de la ulexita y dióxido de carbono. Bolivian Journal of Chemistry. 2013; **30**(1):42-49

[14] Panorama de Mercado de Rocas y Minerales Industriales. Boratos. Ministerio de Energía y Minería. Presidencia de la Nación. Dirección de Asistencia al Productor Minero. Dirección Nacional de Promoción Minera. Argentina: Subsecretaría de Desarrollo Minero; 2018

[15] Chong G, Garcés I. Los yacimientos de boratos en Chile y su beneficio. Colegio de Ingenieros de Chile. 1989; **100**:37-47

[16] Barker J, Lefond S. Borates: Economic Geology and production. Society of Mining Engineers of the

American Institute of Mining: Metallurgical and Petroleum Engineers, Inc.; 1985. p. 274

[17] Harben PW, Kuzvart M. Industrial Minerals, a Global Geology. Industrial Minerals Information Ltd.; 1997. p. 462

[18] USGS. Mineral Industry Surveys. Tables-only release. 2017. Available from: https://www.usgs.gov/centers/nmic/boron-statistics-and-information Tabla datos

[19] Garcés IM. Modelización geoquímica de soluciones concentradas: Aplicación al estudio de la evolución de algunos salares tipo chilenos [tesis doctoral]. Zaragoza, España: Universidad de Zaragoza; 2000. p. 225

[20] Munk LA, Hynek SA, Bradley D, Boutt D, Labay K, Jochens H. Lithium brines: A global perspective. Reviews in Economic Geology. 2016;**18**:339-365. Available from: https://pdfs.semanticscholar.org/3616/dd9db704d285b027c13185f4b6af3b9f2262.pdf

[21] Tassara A. Segmentación andina desde el análisis flexural de la anomalía de Bouguer [unpub. MSc thesis]. Santiago de Chile: Universidad de Chile; 1997. p. 140

[22] Garcés I, Chong G. Yacimientos de Boro de Chile. Características, usos, mercados y aplicaciones industriales de los boratos. Rev. Innovación. 1993;**1**: 23-36. ISSN: 0716-6311

[23] Chong G, Garcés I. Antecedentes sobre los recursos económicos de los salares. In: Caracterización geológica química de los salares chilenos tipos y su consecuente importancia económica. Informe inédito. Fondecyt 786/87, Santiago, Chile. 1988

[24] Garcés I. Procesos de concentración de Boratos (ulexita) en depósitos chilenos. In: Proceedings of the 6° Congreso Mediterráneo de Ingeniería Química. EXPOQUIMIA'93. Barcelona, España. 1993. pp. 665-667

[25] Dunai TJ, López GA, Juez-Larré J. Oligocene-Miocene age of aridity in the Atacama Desert revealed by exposure dating of erosion-sensitive landforms. Geology. 2005;**33**(4):321-324. DOI: 10.1130/G21184.1

[26] Hall SR, Farber DL, Audin L, Finkel RC, Mériaux AS. Geochronology of pediment surfaces in southern Peru: Implications for Quaternary deformation of the Andean forearc. Tectonophysics. 2008;**459**(1):186-205. DOI: 10.1016/j.tecto.2007.11.073

[27] Hall SR, Farber DL, Audin L, Finkel RC. Recently active contractile deformation in the forearc of southern Peru. Earth and Planetary Science Letters. 2012;**337**:85-92. DOI: 10.1016/j.epsl.2012.04.007

[28] Kober F, Ivy-Ochs S, Schlunegger F, Baur H, Kubik PW, Wieler R. Denudation rates and a topography-driven rainfall threshold in northern Chile: Multiple cosmogenic nuclide data and sediment yield budgets. Geomorphology. 2007;**83**(1):97-120. DOI: 10.1016/j.geomorph.2006.06.029

[29] Placzek CJ, Matmon A, Granger DE, Quade J, Niedermann S. Evidence for active landscape evolution in the hyperarid Atacama from multiple terrestrial cosmogenic nuclides. Earth and Planetary Science Letters. 2010;**295**(1): 12-20. DOI: 10.1016/j.epsl.2010.03.006

[30] Benavente C, Zerathe S, Audin L, Hall R, Robert X, Delgado F, et al. Active transpressional tectonics in the Andean forearc of southern Peru quantified by 10Be surface exposure dating of an active fault scarp. Tectonics. 2017;**36**. DOI: 10.1002/2017TC004523

[31] Vila T. Geología de los depósitos salinos andinos. Provincia de

Antofagasta, Chile: Instituto de Investigaciones Geológicas; 1975. Available from: http://www.andeangeology.equipu.cl/index.php/revista1/article/viewFile/731/1065

[32] Garcés I. The sodium sulphate and ulexite deposits: Salar of Surire, Chile. In: Geertman RM, editor. Proceedings of 8th World Salt Symposium, Salt'2000. Vol. 2. Amsterdam: Elsevier; 2000. pp. 1159-1160

[33] Munk LA, Boutt D, Hynek S, Moran B. Hydrogeochemical fluxes and processes contributing to the formation of lithium-enriched brines in a hyper-arid continental basin. Chemical Geology. 2018;**493**:37-57. DOI: 10.1016/j.chemgeo.2018.05.013

[34] Risacher F, Alonso H, Salazar C. The origin of brines and salts in Chilean salars: A hydrochemical review. Earth-Science Reviews. 2003;**63**(3):249-293. DOI: 10.1016/S0012-8252(03)00037-0

[35] Garcés I. Aplicación del Modelo de Interacción Iónica de Pitzer para determinar Coeficientes de Actividad en Salmueras Naturales con Presencia de Boro. Información Tecnológica. 2019; **30**(2):283-292. DOI: 10.4067/s0718-07642019000200283

[36] Risacher F, Alonso H, Salazar C. Geoquímica de aguas en cuencas cerradas: I, II, III regiones—Chile. Volumen I, II, III. Convenio de Cooperación DGA–UCN–IRD. S.I.T. N° 51. Santiago de Chile; 1999

[37] Li J, Wang F, Michalski G, Wilkins B. Atmospheric deposition across the Atacama Desert, Chile: Compositions, source distributions, and interannual comparisons. Chemical Geology. 2019;**525**:435-446. DOI: 10.1016/j.chemgeo.2019.07.037

[38] Warren JK. Non-potash salts: Borates, Na-sulphates, Na-carbonate, lithium salts, gypsum, halite and zolites. In: Evaporites. Cham: Springer; 2016. pp. 1187-1302. ISBN: 978-3-319-13512-0

[39] Risacher F, Fritz B. Origin of salts and brine evolution of Bolivian and Chilean salars. Aquatic Geochemistry. 2009;**15**(1–2):123-157. DOI: 10.1007/s10498-008-9056-x

[40] Vila T. Geología de los depósitos salinos andinos. Provincia de Antofagasta [Tesis de Ciencias Geológicas]. Santiago de Chile: Universidad de Chile; 1974

[41] Alpers CN, Wittermore DO. Hydrogeochemistry and stable isotopes of ground and surface waters from two adjacent closed basins, Atacama Desert, northern Chile. Applied Geochemistry. 1990;**5**:719-734

[42] Gannat E, Schlund JM. Informe de la misión de estudio de las posibilidades de Chile en minerales potásicos. Informe Inédito, vol. 2 Santiago, Corp. Pom. Produc. y Bur. Rech. Geol. Min.; 1969

[43] Stoertz GE, Ericksen GE. Geology of salars in northern Chile. In: U.S. Geol. Surv. Prof. Paper 811. 1974. p. 65

[44] Ericksen GE, Salas OR. Geology and resources of salars in the Central Andes. In: Ericksen GE, Cañas PM, Reinemund JA, editors. Geology of the Andes and Its Relation to Hydrocarbon and Mineral Resources, Series. Vol. 11. Houston, Texas: Circum-Pacific Council for Energy and Mineral Resources; 1989. pp. 151-164

[45] Wisniak J, Garcés I. The rise and fall of the salitre (sodium nitrate) industry. Indian Journal of Chemical Technology. 2001;**8**(5):427-438. ISSN: 0971-457X (Print)

[46] Dirección General de Aguas (DGA). Balance Hídrico de Chile. Gobierno de Chile: Ministerio de Obras Públicas,

Transporte y Telecomunicaciones; 1987. p. 24

[47] Garcés I. Salar de Surire un ecosistema altoandino en peligro, frente a escenario del cambio climático. Rev. Nexo. 2011;**24**(1):43-49

[48] Garcés I, López P. Antecedentes Hidroquímicos del Salar de Aguas Calientes I (Chile). Revista Facultad de Ingeniería Universidad de Antioquia. 2012;**62**:91-102

[49] Lindsay J, de Silva S, Trumbull R, Emmermann R, Wemmer K. La Pacana caldera, N. Chile: A reevaluation of the stratigraphy and volcanology of one of the world's largest resurgent calderas. Journal of Volcanology and Geothermal Research. 2001;**106**:145-173. DOI: 10.1016/S0377-0273(00)00270-5

[50] Alonso R. On the origin of La Puna borates. Acta Geológica Hispánica. 1999; **34**:141-166

[51] Harvie C, Weare J-H. The prediction of mineral solubilities in natural waters: The Na-K-Mg-Ca-Cl-SO_4-H_2O system from zero to high concentration at 25°C. Geochimica et Cosmochimica Acta. 1980;**44**:981-997. DOI: 10.1016/0016-7037(84)90098-X

[52] Harvie C, Eugster H, Weare J. Mineral equilibria in the six-component seawater system, Na-K-Mg-Ca-SO_4-Cl-H_2O at 25°C. II: compositions of the saturated solutions. Geochimica et Cosmochimica Acta. 1982;**46**:1603-1618. DOI: 10.1016/0016-7037(82)90317-9

[53] Harvie CE, Moller N, Weare J. The prediction of mineral solubilities in natural waters: The Na-K-Mg-Ca-H-Cl-SO_4-OH-HCO_3-CO_3-CO_2-H_2O system to high ionic strengths at 25°C. Geochimica et Cosmochimica Acta. 1984;**48**:723-751. DOI: 10.1016/0016-7037(84)90098-X

[54] Marion GM, Farren RE. Mineral solubilities in the Na-K-Mg-Ca-Cl-SO_4-H_2O system: A re-evaluation of the sulfate chemistry in the Spencer–Møller–Weare model. Geochimica et Cosmochimica Acta. 1999;**63**(9): 1305-1318. DOI: 10.1016/S0016-7037(99)00102-7

[55] Pabalan RT, Pitzer K. Thermodynamics of concentrated electrolyte mixtures and the prediction of mineral solubilities to high temperatures for mixtures in the system Na-K-Mg-Cl-SO_4-OH-H_2O. Geochimica et Cosmochimica Acta. 1987;**51**: 2429-2443. DOI: 10.1016/0016-7037(87)90295-X

[56] Spencer R, Moller N, Weare J. The prediction of mineral solubilities in natural waters: A chemical equilibrium model for the Na-K-Ca-Mg-Cl-SO_4-H_2O system at temperatures below 25°C. Geochimica et Cosmochimica Acta. 1990;**54**:575-590. DOI: 10.1016/0016-7037(90)90354-N

[57] Plummer L, Parkhurst DL, Fleming G, Dunkle S. Computer program incorporating Pitzer's equations for calculation of geochemical reactions in brines. In: Water-Resources Investigation Report. WRI 88-4153. 1988. p. 310

[58] Castellán GW. Fisicoquímica. Pearson Educación; 1987. ISBN: 978-96-8444-316-7

[59] Jander G, Friedrich JK. Análisis Volumétrico. Unión Tipográfica Editorial Hispano Americana; 1961. p. 400

[60] Pickering W. Química analítica moderna. Reverté S.A. 1a edición. España; 2011. p. 688

[61] Ball JW, Nordstrom DK. User's manual for WATEQ4F, with revised thermodynamic data base and test cases for calculating speciation of major, trace, and redox elements in natural waters. In: U.S. Geological Survey Open-File Report 91-183. 1991. p. 189

[62] Jenne EA, Ball JW, Burchard JM, Vivit DV, Barks JH. Geochemical modeling: Apparent solubility controls on Ba, Zn, Cd, Pb and F in waters of the Missouri Tri-State mining area. In: Hemphill DD, editor. Trace Substances in Environmental Health. Vol. 14. Columbia, Montana: University of Missouri; 1980. pp. 353-361

[63] Dickson AG. pH scales and proton-transfer reactions in saline media such as sea water. Geochimica et Cosmochimica Acta. 1984;**48**: 2299-2308. DOI: 10.1016/0016-7037(84)90225-4

[64] Markus Y. Determination of pH in highly saline waters. Pure and Applied Chemistry. 1989;**61**:1133-1138. DOI: 10.1351/pac198961061133

[65] Lazar B, Starinski A, Katz A, Sass E, Ben-Yaakov S. The carbonate system in hypersaline solutions: alkalinity and $CaCO_3$ solubility of evaporated seawater. Limnology and Oceanography. 1983;**28**:978-986. DOI: 10.4319/lo.1983.28.5.0978

[66] López PL, Garcés I, Auqué LF, Gómez J, Gimeno MJ, Lapuente MP. Características geoquímicas y aspectos geológicos del salar de Huasco (Chile). Boletín Geológico y Minero. 1999;**110**: 297-316

[67] Pengsheng S, Yan Y. Thermodynamics and phase diagram of the Salt Lake brine system at 298.15 K: V. Model for the system Li^+, Na^+, K^+, Mg^{2+}/Cl^-, SO_4^{2-}, H_2O and its applications. Calphad. 2003;**27**(4): 343-352. DOI: 10.1016/j.calphad.2004.02.001

[68] Gao F, Zheng M, Song P, Bu L, Wang Y. The 273.15-K-isothermal evaporation experiment of lithium brine from the Zhabei Salt Lake, Tibet, and its geochemical significance. Aquatic Geochemistry. 2012;**18**(4):343-356. DOI: 10.1007/s10498-012-9168-1

[69] Hardie LA, Eugster HP. The evolution of closed-basin brines. Mineralogical Society of America Special Paper. 1970;**3**:273-290

Chapter 2

Functionalized Boron Nitride Applications in Biotechnology

Hélio Ribeiro, Paula von Cardoso Randow, Diego N. Vilela, Milene Adriane Luciano and Lidia Maria de Andrade

Abstract

Due to its interesting chemical, physical, and biological properties, boron nitride has received considerable attention by the scientific and technological communities. However, there is a strong dependency of its structural quality and compatibility in different host systems, regarding its potential applications. The use of these different nanostructures involves several challenges due to their low dispersibility in water and organic solvents; thus, its chemical modification is an important step that gives them specificity. Therefore, the ability to control their surface (physically or chemically) is essential for exploring and building blocks in the nanoengineering of supramolecular structures. In this chapter, we report different boron nitride functionalization processes, as well as their important uses as adjuvants in vaccines, brachytherapy, or drug delivery. Besides some important theoretical studies that have demonstrated the different functionalization possibilities for use in nanomedicine, are also reported.

Keywords: hexagonal boron nitride, boron nitride nanotubes, functionalization, nanomedicine, cancer, drugs

1. Introduction

Boron nitride (BN) is a covalent solid constituted by equal numbers of boron and nitrogen atoms, and its well-known crystalline forms are hexagonal boron nitride (h-BN), diamond-like cubic BN (c-BN), and wurtzite BN (w-BN) [1]. In particular, h-BN is the most stable BN phase under standard conditions that presents two-dimensional (2D) layered structure. The sp^2 hybridization is observed for boron atoms, and bond angle expected between the B-N-B or N-B-N is 120° that forms a perfect hexagonal ring bond network such as graphite, whereas the 2D layers are held together by weak van der Waals forces [1, 2]. Due to its similar structure with graphite, h-BN has common properties with it, like anisotropy along and perpendicular to a basal plane, high mechanical strength and thermal conductivity, and excellent lubrication [1]. Due to its reduced electron delocalization in the π BN bonds, it has a large bandgap making it an electrical insulating nanomaterial; however, h-BN is used as a lubricant (weakly held layers can slide over each other) and can have semiconductor properties. h-BN is also highly thermally and chemically stable and thus is also widely used for durable high-temperature crucibles, antioxidation lubricants, and protective coatings [1] and also has a wide range application in nanomedicine and biotechnology such as cancer therapies [3],

IntechOpen

drug delivery [4, 5], biosensing [6–8], adjuvant in vaccines [9], and intracellular delivery [10, 11]. Hexagonal boron nitride also can be incorporated in ceramics, alloys, resins, plastics, and rubbers to give them thermal and chemical stability and mechanical resistance as well [12–15]. BN nanomaterials (**Figures 1** and **2**), such as carbon nanotubes (CNT) and graphene, also are awakened as new perspectives in prophylactic, diagnostic, and therapeutic areas. However, the use of them depends strongly of its chemical and physical modification surfaces that display an important role in their compatibility different host environments.

Its modification surfaces results in homogeneous dispersion medium with minimum restacking or agglomeration which guarantees biocompatibility in different biological or inorganic systems. The nanostructure dispersions in different medium of interest are fundamental to their potential applications [16]. The nanoengineering interfaces between host (organic or inorganic) systems and these nanostructures involve several challenges that need to be overcome. For instance, the restacking or agglomeration process of them does not allow transferring their expected properties to the host system, resulting in an inhomogeneous medium with minimum of compatibility. This undesirable phenomenon of surface can be overcome by physical or chemical modification methodologies, such as covalent or non-covalent functionalization. Thus, our choices will depend on the nanostructure and system desired. There are several types of covalent functionalization, and different functional groups can be chosen, according to their selectivity [16]. In this case, different organic or inorganic functional groups or nanoparticles can be anchored by strong covalent bonds. For instance, it can be introduced on surfaces of oxidized CNTs or graphene oxide (GO), functional groups such as alkoxy (–OR), amino (–NH_2), amine (–NHR), alkyl (–R) [17, 18], heteroatom doping, metallic nanoparticles, biomolecules, and biopolymers, among others. These modification

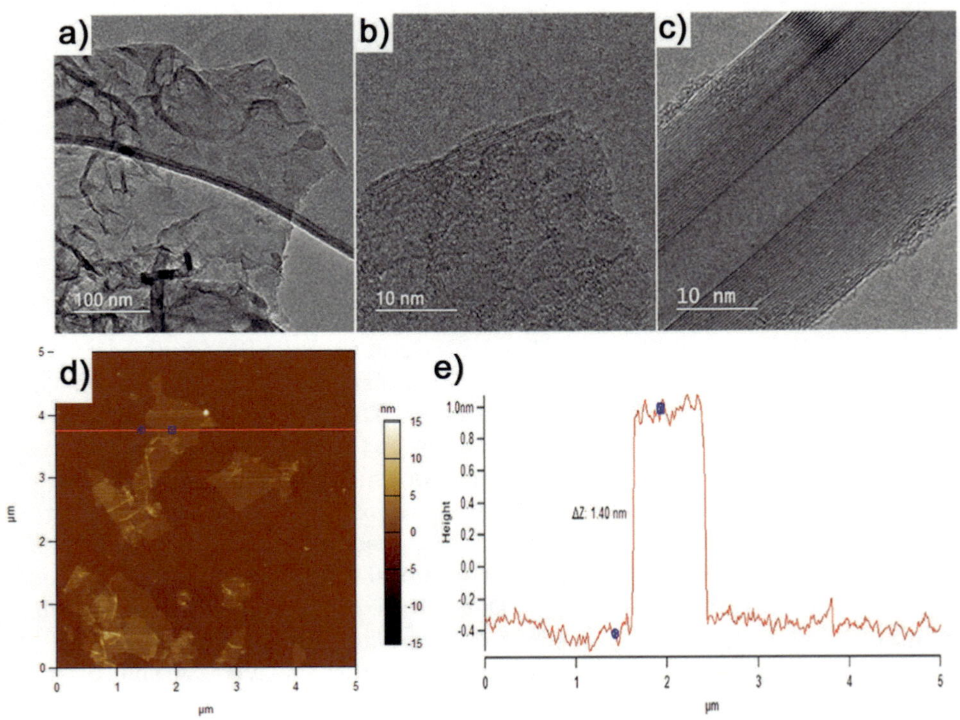

Figure 1.
TEM images of (a, b) h-BN in different magnitudes, (c) multilayered BNTs. (d, e) AFM image and profile of h-BN nanosheets. (Images courtesy of the UFMG Microscopy Center).

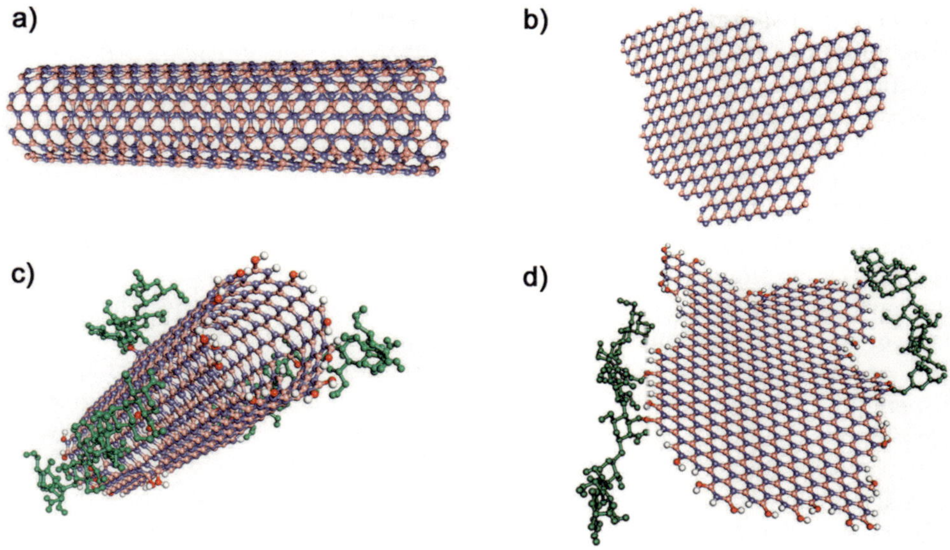

Figure 2.
BNNT and h-BN scheme not functionalized (a, b) and functionalized with different functional groups (c, d).

processes alter significantly their surface interactions, leading them to a large range of solubility in water, copolymers, or organic solvents [16].

On the other hand, non-covalent functionalization of nanoparticles is strongly dependent on their physical interaction with host system through intermolecular forces, such as van der Waals, hydrophilic, hydrophobic, hydrogen bonding, and π-π interactions, among others [19]. Using advantages of these physical interactions of molecules (conjugated, surfactants, etc.), they form homogenously dispersion into different medium with controlled physicochemical properties [20]. The study of different modification surface and their potential applications is very important. Several studies have presented different types of covalent and non-covalent functionalization methodologies in different nanoparticles with several demands, such as in cellulose films or fibers [21, 22], chitosan [23], polyethylenimine-grafted nanoribbon (for recognition of microRNA) [8], vinyl acetate copolymer [20], octadecylamine [1], glucose oxidase biosensing [24], poly (ethylene glycol) [25], DNA [26], and metallic nanoparticles, among others. The most important issue for success and performance of these compounds is the choice of the best tailored functionalization process (bio-nanoengineering) for each type of inorganic or biological system. The physical-chemical modification is an essential step for their relevant applications, leading to nanocomplex chemically stable, well dispersed, and biocompatible with the biological environment of interest. Thus, it is possible to produce smart nano-systems with advanced applications in biotechnology and biomedical areas, such as ecological packaging, bio-robots, and tumor markers for diagnosis and therapy. In this chapter, we propose a brief report about the main methods of functionalization of boron nitride and their advances and potential applications in biotechnology and nanomedicine.

2. Boron nitride functionalization and its implications in nanomedicine

Hexagonal boron nitride is thermal and chemically layered structure more stable than graphitic carbon structures. Due to its excellent chemical stability, a better biological performance is expected [27]. The key for application of h-BN in different systems is its dispersion by the introduction of specific agents (such as surfactants) or by some covalent attachments [28]. The adequate modification process provides specific surfaces that allow its better solubility or dispersion in different solvents media playing an important role regarding selectivity, for instance, in biological systems [28]. These nanoparticles can be coated with biological molecules to facilitate their interaction with a living system, binding to a specific target, like tumor cells [29]. Regarding the tumor targeting strategies, the enhanced permeability and retention (EPR) effect of nanomaterials is a key mechanism for solid tumor targeting, and it is considered a gold standard for novel radioisotopes design [30]. There are two different types of possible modifications. One is to attach, covalently, a molecule or molecular structure, and the second approach is through physical adsorption onto BN or h-BN surface nanostructure. In addition to improve solubility and dispersion, an important issue that needs to be addressed is the toxicity of boron nitride materials and their behavior in a biological system. The primary objective of toxicology studies is to evaluate the safety of potential drug candidates, using animal models and validated procedures to access genetic toxicology, acute and sub-chronic toxicology, and absorption, distribution, metabolism, and excretion (ADME) studies [31]. The ability to design, format, and run assays that are specific, sensitive, and robust is crucial in biomedical research and plays an important role in launching a drug development project to a successful outcome. One of the most important prerequisites that a new drug must fulfill is nontoxicity. Cell culture can be used to access toxicity and cytocompatibility of BN, detecting several parameters such cytosolic enzyme release, cell growth, and cloning efficiency. Many factors can be defined directly related to chemical kinetics like absorption rates, membrane permeability, intracellular synthetic pathways, biotransformation, oxidative stress, or apoptosis/necrosis induction, distribution, and excretion [32, 33]. The modulation of immune cell response can be evaluated at gene and protein levels, mainly regarding the expression of cytokines and chemokines [33].

A study showed that important cytotoxic effects of pure BNNTs on several cell lines were observed with a concentration of 20 μg/mL. According to this viability test, the levels of toxicity and morphological alterations were cell-type dependent and related to the absence of a biomolecule, causing acute toxicity in cell cultures, due to the increased nanomaterial accumulation [34].

In an interesting study, Ciofani et al. [10] proposed a technique to obtain BNNT stable dispersions for biological applications, dissolving BNNT powder in polyethylenimine (PEI) water solutions via an ultrasonication process. PEI is a cationic polymer for nucleic acid delivery with a high transfection efficiency, enhanced cellular uptake, and endosomal escape. In this in vitro testing performed on human neuroblastoma cell line (SH-SY5Y), PEI-coated BNNTs showed good cell viability in the culture medium. Another study demonstrated that pectin-coated BNNTs (P-BNNTs) are nontoxic for macrophages up to 50 μg/mL after 24 h of incubation. According to this study, cytokine expression was not affected by the administration of the P-BNNT, and its uptake by macrophages did not cause any cell membrane impairment, adverse effects, or inflammation processes in the cell [33]. BN coated with glycol-chitosan (GC) assay was performed to evaluate the BNNT interactions with biological systems and determine their biosafety, using human vein endothelial cells (HUVECs). Various parameters were observed such as cell toxicity, proliferation, cytoskeleton integrity, cell activation, and DNA damage. At the highest concentration, only a small reduction in cell viability and the increase of a vascular adhesion molecule (a marker of cell activation) expression were identified. According to that study, these findings show that GC-BNNTs do not affect endothelial cell biology and can be further investigated for vascular targeting, imaging, and drug delivery [35].

Cytosine-phosphate-guanine (CpG) oligodeoxynucleotides (ODNs) are molecules that present promising therapeutic properties due to their capability of stimulating innate and adaptive immune responses, activating Toll-like receptor 9 (TRL-9), leading to induction of proinflammatory cytokines [36]. The CpG ODNs delivery system was developed by the functionalization of boron nitride nanospheres with PEI. The complex BNNs-PEI showed enhanced dispersity and stability in aqueous solution. Although PEI itself can be used as a drug delivery carrier, the complex BNNs-PEI exhibited much higher efficiency than the PEI alone, increased CpG ODNs' cellular uptake and induced higher expression of interleukin-6 (IL-6) and tumor necrosis factor-α (TNF- α) from peripheral blood mononuclear cells. No obvious cytotoxicity was observed with concentrations up to 100 μg/mL [35].

It was demonstrated that the complex PEI-BNNTs is internalized through an energy-dependent process via endocytosis pathway. PEI coating influences the behavior of BNNTs as it enhances the chemical reactivity of the nanotubes, allowing the interaction with a biological system [37]. In another cell viability assay [36], the MRC-5 culture, a diploid cell line derived from human fetal lung, normally used as a model for early cytotoxicity studies was treated for 48 h with samarium doped boron nitride nanotubes (Sm- BNNTs). This study demonstrated that even at the highest concentration, the cell viability was higher than 70%. This result reassures the findings of previous studies [29, 37, 38], confirming that, in concentrations below 200 μg/mL, BNNTs showed low cytotoxicity, and BNNTs do not appear to inhibit cell growth or induce apoptotic pathways in the cells. It should be noted that the high purity and quality of BNNTs are crucial for their nontoxicity [11, 36]. Many studies were conducted to determine possible toxicity to metabolism and liver morphology, as well as to renal function. A biodistribution study shows a significant elimination of BNNTs by renal excretion and accumulation in the liver, spleen, and intestines [36]. Ciofani et al. results indicate a similar pharmacokinetic behavior in an in vivo investigation of BNNTs in rabbits, with relatively high clearance of GC-BNNTs from the blood, with quick distribution and excretion [39]. The hematological parameters were monitored for up to 7 days. During the observation period, neither unusual behavior including sweating, excitement, trembling, head nodding, or temperature changes was observed. The study showed excellent hematological results without evidence of adverse effect, liver or kidney impairment [18]. A biocompatibility study with planarians indicated that gum Arabic-coated BNNTs do not induce oxidative DNA damage and apoptosis and did not show adverse effects on animal stem cell biology or on tissue regeneration [40].

As the cytotoxicity and biocompatibility of nanomaterials are being elucidated, new approaches overcome this first phase of research, leading to new discoveries and opening doors to novel alternatives. The antimicrobial activity of nanoparticles was investigated on different microorganisms in a study with h-BN. It was demonstrated that the h-BN has a bacteriostatic effect and shows a high antibiofilm activity on preformed biofilm [38]. PEI-BNNTs and other surfactant-coated BNNTs were also examined to elucidate their antibacterial activity. The optical density of bacterial growth curves and the transmission electron microscopy morphology images revealed that PEI-BNNTs exhibit strong microbial activity against *Escherichia coli* and *Staphylococcus aureus* [41]. Because of its low toxicity, high mechanical strength, and chemical stability, BNNTs also are considered to be a promising bioactive material for bone tissue engineering, improving polymers, composites, and scaffold properties [42]. In a study, akermanite (AKM), a bioactive material, was reinforced with boron nitride nanosheets (BNNSs) to ameliorate its mechanical features [43]. Significantly enhancement in compressive strength, fracture toughness, and an overall favorable cytocompatibility characteristics were found. Analysis of osteoblast gene expression and alkaline phosphatase activity measurement suggest that nanomaterials have a great potential in the orthopedic implant field [28]. Chitosan-based scaffolds are thoroughly studied owing to their biocompatibility and antimicrobial activity. Hydroxylated BNNTs (BNNT-OH) were included into a chitosan scaffold and tested for their mechanical strength. The results of this study indicated that the inclusion of BNNT-OH increased mechanical strength and induced cell proliferation and adhesion decreased the scaffold degradation rate when compared to chitosan-only scaffold and did not cause toxicity to human dermal fibroblast cells [42]. Despite of some different experimental approaches and nanomaterial complexity, most of the studies show very good response of cells and organisms toward boron nitride nanocompounds. Researches about boron nitride applications in biotechnology are at full expansion, and nanomaterial biocompatibility and biosafety require further investigation.

2.1 Functionalized BN as a candidate for imaging and cancer therapy

Nowadays, some challenges regarding the earliest tumor diagnosis and treatment have been discussed considering that time represents a critical point between death and surviving. In this way, nanotechnology appears as promising tool due to its small size and remarkable physical-chemical characteristics as well, opening the as-called nanomedicine era. Traditional imaging resources to identify tumor sites have been applying ionizing radiation equipment. One of the most used is computed tomography followed by scintigraphy such as positron emission tomography coupled with computed tomography images (PET-CT). Likewise, non-ionizing radiation tools are applied such as magnetic resonance image (MRI). However, even though all these strategies possess high-image resolution, they are not completely able to detect smaller precocious tumors a primary indicative of micrometastasis.

There are several efforts to enhance the quality of images especially those taken from MRI. Notwithstanding, technical advances in nanotechnology are creating novel classes of MRI contrast-enhancing agents offering much higher relaxivities than most current clinical contrast agents, which translates into greater MRI contrast enhancement. These nanoscale agents also have the potential to revolutionize in vivo applications of contrast-enhanced MRI since they offer the multiple advantages of low toxicities, extremely high relaxivities, and cell internalization capabilities [44]. One of these nanomaterials presenting good possibilities to enhance MRI contrast is boron nitride nanotubes (BNNTs) containing Fe paramagnetic impurities. One drawback of pristine BNNT is its high hydrophobicity, and to overcome the low solubility of pristine BNNTs in aqueous solution, they can be wrapped with poly-L-lysine creating a PLL-BNNT. This soluble BNNT containing Fe has demonstrated values of transverse relaxivities comparable to commercial superparamagnetic iron oxide nanoparticles suggesting Fe-BNNT as a potential magnetic-enhanced contrast agent for MRI images at a field of 3 T [45]. Much has been investigated to improve the beneficial effects of radiotherapy especially in that case where radioresistant behavior is observed [46]. Radiotherapy is one of the most largely and effective tools to treat cancer using different approaches among then ^{10}B through the boron neutron capture therapy (BNCT). The fundamental concept of boron neutron capture therapy is the production of high-linear energy transfer (LET) particles when one “tags” or “labels” a tumor cell with a compound having a large cross section capable of capturing a neutron. After neutron capture it goes into excited state releasing local energy driving heavy ion products over short distances comparable to the dimensions of cells ~20 μm [47]. The BNCT reaction is presented in **Figure 3**.

BNNTs have been thought as a possible candidate to BNCT due to their considerable high B density. Preliminary in vitro studies have suggested

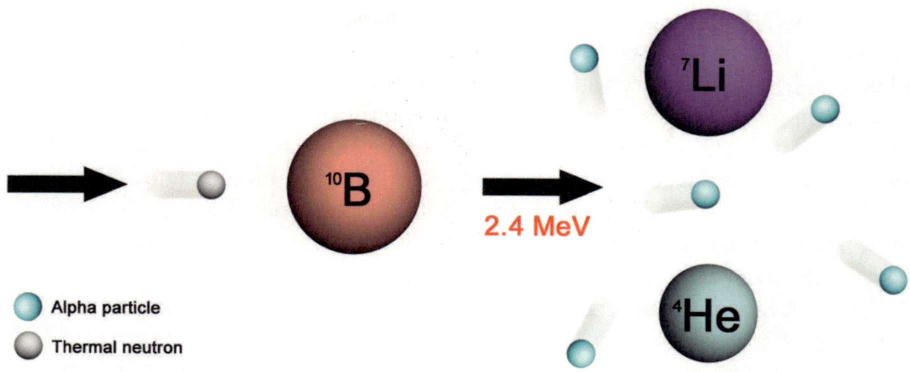

Figure 3.
BNCT reaction scheme. ^{10}B bombarded by thermal neutrons resulting in a nuclear reaction where ^{7}Li nuclei and α particles are released offering a lethal effect localized within a tumor cell.

folate-functionalized BNNTs (FA-BNNT) as a potential boron delivery agent to malignant glioblastoma cells whose results have confirmed a strong and selective uptake of these nanotube vectors by human glioblastoma multiforme T98G cells and for cervix tumor cell line HeLa, but not by normal human fibroblasts [48, 49]. PEGylated BNNT was evaluated as a BNCT strategy in B16 melanoma cell line compared with the conventional B carrier sodium borocaptate (BSH). Although PEGylated BNNT have shown three times higher cellular uptake than BSH, the concentration of ^{10}B delivered wasn't enough to perform BNCT as well as gold nanoparticles functionalized with carboranes as a ^{10}B donator. Even though these nanoparticles have demonstrated high accumulation of B atoms inside cells, they have failure in therapeutic window for BNCT, then requiring improvements in their design as a ^{10}B source [50, 51].

Other applications of BNs in the field of cancer treatment due to their physical properties have been proposed. One of them presents a new boron-containing chlorin derivative as an agent for photodynamic therapy (PDT) and BNCT. Using in vivo xenographic tumor model, the results have shown significant tumor growth inhibition suggesting this nanocompound as a promising agent for PDT and for BNCT as well [52]. Radioisotopes can be used as image resource in nuclear medicine and in some cases as a therapy agent too. BNNT doped with samarium 152, a radioactive isotope from a rare earth metal of the lanthanide group, has shown low toxicity in MRC-5 fibroblasts suggesting this nanomaterial as a potential nanosized β-emission source for nuclear medicine therapy especially for bone metastasis treatment [30].

Another widely used pathway to cancer treatment is chemotherapy. Despite efficient outcomes of chemotherapeutical agents, side effects are usual, and, in some cases, *enhanced permeability and retention* (*EPR*) *effect is hampered. Doxorubicin* (*DOX*) *and* paclitaxel *are a broad-spectrum antitumor drugs widely used for the treatment of several kinds of cancers, including prostate, breast and ovarian. As an attempt to improve chemotherapy effects*, different types of nanocarriers are under investigation especially as selectively drug deliver, commonly using liposomes and carbon nanotubes (CNTs). The utility of BNNTs@NaGdF4:Eu composites as a targeted cancer therapeutic for chemotherapy drug delivery, used to load doxorubicin in the absence and presence of a magnetic field, has been demonstrated. By using human prostate cancer LNCaP cell line, the BNNTs@NaGdF4:Eu composite shows a loading efficiency of doxorubicin of about 30% when a magnetic field is applied [53]. Besides, the administration of hollow BN spheres with paclitaxel leads to synergetic effects in the suppression of tumor growth of human prostate cancer as in vitro as in vivo mainly due to apoptosis death [54].

Despite BNNTs possess extraordinary physical-chemical properties highlighting the potential applications of these nanomaterials in nanomedicine, they have been poorly exploited [55, 56]. The safety and cytobiocompatibility of boron nitride compounds have been reported in vitro using glioblastoma multiforme T98G cells [48], human MRC-5 fibroblast cells [48], human cervix tumor cell line HeLa [49], melanoma B16 cells [50], rat-derived osteosarcoma cell line UMR-106 [51], human neuroblastoma cells SH-SY5Y [37, 57], and human embryonic kidney cell line HEK 293 [11]. However, based on the fact that nanotubes present some similarities with asbestos lung cancer, human lung cancer cells (A549), alveolar macrophages (RAW264.7), fibroblast cells (3T3-L1), and human embryonic kidney cells (HEK293) were also exposed to BNNTs to verify cytotoxicity regarding BN compounds. In fact, the results indicate that BNNTs are cytotoxic in vitro even at low concentration in a time and dose-dependent manner promoting morphological alterations and decreased cellular viability [34].

Nanomaterials applied in medicine are a controversial issue requiring further investigations, especially due to a potential toxicity and environmental consequences regarding B compound disposal, and their biological responses need to be examined like extensively related CNT cytotoxicity reports. Recent progress in the field of nanomedicine research has offered good alternatives on this matter, especially those targeting tumor cells with metal nanoparticles [58]. For a while, the insipient results highlight a promising future for nanomedicine leading to new therapeutic approaches. For instance, graphene oxide may be an effective nontoxic therapeutic strategy for the eradication of cancer stem cells, via differentiation-based nanotherapy [59]. As described above, boron nitride represents one of the most promising nanomaterials in nanomedicine since some drawbacks can be overcome such as improvement of solubility in aqueous solution, high ^{10}B concentrations releasing, allowing their use for BNCT and reducing of cytotoxicity. Moreover, further preclinical investigations will provide the safety and efficacy of BN for image and therapy leading to a more deeply knowledge about biodistribution, clearance, and side effects regarding dose, time, and administration pathways.

3. Computational simulation perspectives of boron nitride nanotubes using as drug carriers for cancer treatment

The advances in nanotechnology have stimulated a large production and storage of information coming from different sources, such as high-processing DNA sequencers, nuclear magnetic resonance, X-ray crystallography, electron microscopy images, and spectroscopic data, among others [60]. This large number of information about different systems needs to be quickly processed and understood in terms of structure, morphology function, and properties being solved by computational simulation (CS) with efficiency and high degree of accuracy [60]. CS has been widely used in studies of aggregation, folding, functionality, and nanoengineering platforms for several proteins, tertiary structures of RNA, and nanostructures surfaces and arises in the development of new techniques and diagnosis and therapy [61]. Thermodynamic properties such as enthalpy, entropy, and free energy of protein conformations also have been used to predict different protein and DNA structures [62], with atomic details applied in complex organisms [63]. By molecular dynamic simulation (MDS), several researchers have investigated the behavior of different nanoparticles, clusters, biomolecules, polymers, ceramics, metal alloys, composites, and fuel systems in relative short-time operation, compared to other non-computational methods. Numerical simulations based on preestablished mathematical models that involve quantum mechanics, molecular mechanics, Monte-Carlo, or a hybrid thereof have been used as well [64].

Quantum mechanics provides the accurate description of the energy about system electronic distribution. However, this description may have some restrictions, such as the size of the system (< 500 atoms) and requires high computational power, due to the large correlation in many electron interactions [65]. This limitation can be overcome by the density functional theory (DFT), to solve many body systems, for example, ionization energy calculations and band theory analysis. In this approach is considered an ab initio method because it does not require initial parameterizations, making this technique faster when used electronic density approximations are used, such as Car-Parrinello and Born-Oppenheimer methods [66]. In the case of molecular mechanics, the potential energy is described in geometric terms by force fields inherent to the molecules, in their internal and external interactions [64]. This approach acts empirically on the integration of potential energies resulted from the intramolecular interactions and intermolecular where the energy of each components or entity must be explicitly parameterized according to the specific equilibrium between atoms and their geometry imposed by their structures [65]. The Monte-Carlo method provides spatial conformation based on the atoms' distribution probabilities predicted by the Boltzmann equilibrium in each random Cartesian coordinate. The selected conformation positions within a tridimensional space, regarding to its rotation, torsion, or twist which needs to be lower than its initial conformation in study [65]. About the computational methods used in nanotechnology, we can highlight several studies in different nanostructures, such as h-BN and BNNTs, explored in several technology areas such as DNA sequencing [66–70], water treatment [71–74], piezoelectric properties [75, 76], and drug delivery [77–79], among others. For instance, the cisplatin (CP) is a well-known agent chemotherapeutic that acts directly on DNA causing defects in the cell repair mechanism and leading to apoptosis. This drug is widely used in treatment of several types of cancer, such as lymphomas, sarcomas, and carcinomas, as well as in the treatment of other types of cancer that are affecting the head and neck, bladder, lung, ovaries, and testicles [80]. However, its cellular uptake sometime is accompanied by large side effects and development of resistance, causing allergic reactions, low immunity to infections, hemorrhages, kidney problems, and gastrointestinal disorders among other effects [80]. Similarly, the doxorubicin is a type of anthracycline that acts in the DNA structure preventing its replication, with important role in relation to other chemotherapeutics that present restrictions, such as cytotoxicity, accumulating irreversibly and compromising muscular tissues, mainly cardiac [81]. The expectation about the use of nanocarriers is related to the possibility whether they eliminate or reduce the harm effects in relation to conventional drugs [82]. From here, we will explore some interesting works about MDS applications considering the potential drug carrier as vector in cancer treatments.

For instance, Khatti and Hashemianzadeh focused their efforts on the diffusion process of water in loading of drugs inside the BNNT [83]. Their study was about single-walled zigzag BNNT (18,0) sample with length 40 Å and diameter 14 Å to represent the system. They use BNNT containing 18 hydroxyl groups (-OH) at the end and another one with saturated -OH groups at the same position. The model systems were solvated in an aqueous solution with a TIP3P in octagonal box over 12 Å. All systems were solvated with 7544 water molecules in total [83]. The simulations were performed with different combinations that also include the carboplatin (CPt) molecule. The MDS studies were carried out by the AMBER 12 package with SANDER module (at 300 K and 1 atm) using SHAKE algorithm, where each simulation included 5000 steps of solvent/solute running through 10 ns to the step of 2 fs [83]. The diffusion coefficient (D) results can be seen in **Table 1**.

The results showed that the electrostatic interaction -OH groups in BNNT cause different diffusion behaviors in water compared to BNNT nonfunctionalized.

System	D (cm^2 s^{-1} × 10^5)
BNNT/water	0.440
BNNT-OH/water	0.600
BNNT-CPt/water	0.560
BNNT-OH/CPt	1.890

Table 1.
Diffusion coefficient (D) prediction of different BNNT systems in different combinations studied by Khatti and Hashemianzadeh (2016) [83].

Therefore, the drug diffusivity into the nanotube can be related to the diffusive behavior of water molecules inside the tube. The increase of the diffusion rate of water molecules in BNNT-OH is observed according to the diffusion coefficient values. These results showed us that the penetration rate of the drug in BNNT-OH is larger than that nonfunctionalized BNNT, and the electrostatic interactions between carboplatin and hydroxyl groups on the tube edge enhance the permeation rate of the drugs into the nanotube cavity [83]. By MSD study, they observed that hydrogen bonds at the extremities of BNNT produce dipole-dipole interactions between hydrophilic CPt and -OH groups, which allow an efficient condition for the encapsulation of this drug. The hydroxyl groups also maintain the stability of the BNNT and prevent its aggregation into the solutions [83]. An encapsulation scheme of CPt molecules encapsulation into BNNT-OH can be visualized in **Figure 4**.

Nanotube structures are of great interest in nanomedicine as a vector in drug delivery; they can access the cell carrying large amounts of drugs due to their possibility endohedral functionalized surfaces [77]. From this juncture, several works have performed different theoretical methods of BNNT stability by drug encapsulation. For instance, the azomethine (AZ) prevents the carboplatin deactivation process, prematurely before the CPt achieves the cancer cell [77]. Studies have demonstrated that BNNT-OH acts as nanocarriers drugs for CPt complex. **Figure 5** shows a scheme of BNNT endohedrally functionalized with carboplatin, and the DNA linked covalently to the CPt complex.

Duverger et al. used DFT simulations to study the interactions between BNNT-OH in azomethine and an anticancer agent (Pt (IV) complex) linked with an amino-derivative chain. The geometry of the azomethine/Pt/amine system was optimized by different molecular configurations on the inner and outer surfaces of the BNNT. The authors also showed that the molecular chemisorption is possible only when the azomethine is present above two adjacent B and N atoms of a hexagon structure. The attachment of an azomethine and subsequent drug did not perturb the cycloaddition process. These theoretical results showed that the therapeutic Pt (IV) complex was not affected when it was attached onto BNNTs [84]. In another work, Khalifi et al. also demonstrated theoretical results about the use of BNNT particles as carrying drug insertion into to the lipid bilayer [85]. They studied a single-walled BNNT armchair (10,10) with diameter 13.94 Å and length 21.32 Å (also saturated with -OH groups) with its extremities encapsulated with carboplatin (CPt) in ionized solvated ambient to mimic a biological environment. The BNNT-CPt complex can move freely, and a bilipid membrane (BM) is formed by 656 molecules of 1-palmitoyl-2-oleoyl-sn-glycero-3-phosphocholine. It was applying DFT method conjugated with classical molecular mechanics [85]. The rapid diffusion of the BNNT-CPt complex toward BM with increased BNNT-CPt hydrophilicity due to surface of BNNT-OH was observed. This diffusion process was energetically favorable to the system due to the contribution of different energy component combinations. According to these results, the possibility of BNNT-OH as a specific ligand for cancer cells allows the more precise encapsulated drug delivery in cell membranes [85].

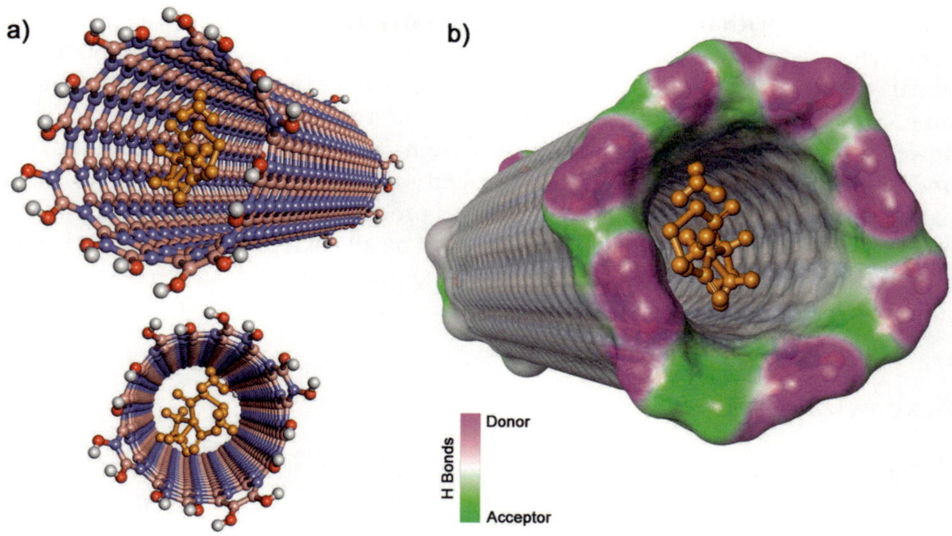

Figure 4.
CPt molecule encapsulated by BNNT-OH (a) and electrostatic force field of BNNT (b).

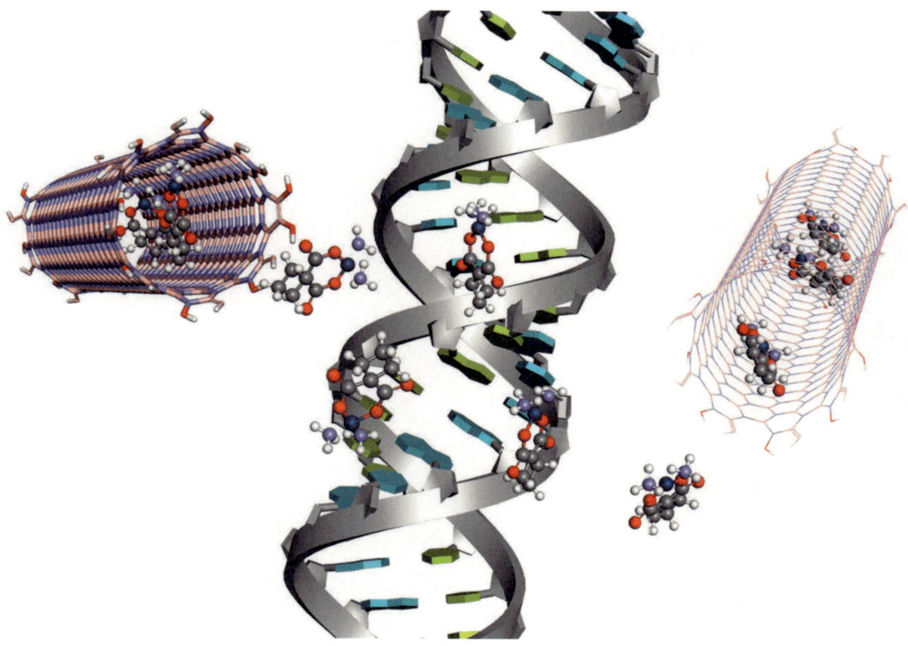

Figure 5.
Theoretical prediction of BNNT (sticks and wires) by delivering CPt molecules directly into DNA, one of the great promises of nanobiotechnology.

4. Conclusion

In this chapter we highlighted some experimental and theoretical works under different BN functionalization structures with potential biomedical applications. However, the most important aspect for success and an optimal performance about these compounds are the choice of the best tailored modification process for each biological system in study. Their physical-chemical modification is an essential step for relevant applications, leading to different systems chemically stable, dispersed, and compatible with the environment of interest. Thus, it is also possible to produce nano-templates with advanced applications in biotechnology. Clearly, computational simulation is an important support tool in science that contributes to reduce time, costs, and risks. In the case of cancer, several types of drugs already exist; however, their delivery to their specific sites remains a big challenge and leads to undesirable consequences. Thus, the computational methods in conjunction with nanotechnology can optimize parameters and procedures maybe improving diagnosis and treatment, the chances of cure. Based on all what have had proposed and investigated, boron nitride arises as a promising nanomaterial for several application even in nanomedicine and nanobiology.

Acknowledgements

The authors thank the researcher Wellington M. Silva (CDTN) and the Microscopy Center of the Federal University of Minas Gerais for the courtesy microscopy images. All images are produced exclusively for this work.

Conflict of interest

The authors also declare that there is no conflict of interest in this work.

Author details

Hélio Ribeiro[1*], Paula von Cardoso Randow[2], Diego N. Vilela[3], Milene Adriane Luciano[1] and Lidia Maria de Andrade[4]

1 Departamento de Química Universidade Federal de Minas Gerais, Belo Horizonte, Brazil

2 Fundação Ezequiel Dias, Belo Horizonte, Brazil

3 Centro Universitário Barão de Mauá, Ribeirão Preto, Brazil

4 Nanobiomedical Research Group, Departamento de Física, Universidade Federal de Minas Gerais, Belo Horizonte, Brazil

*Address all correspondence to: helioribeiro@qui.ufmg.br

IntechOpen

© 2019 The Author(s). Licensee IntechOpen. This chapter is distributed under the terms of the Creative Commons Attribution License (http://creativecommons.org/licenses/by/3.0), which permits unrestricted use, distribution, and reproduction in any medium, provided the original work is properly cited.

References

[1] Weng Q, Wang X, Wang X, Bando Y, Golberg D. Functionalized hexagonal boron nitride nanomaterials: Emerging properties and applications. Chemical Society Reviews. 2016;**45**(14):3989-4012. DOI: 10.1039/C5CS00869G

[2] Taha-Tijerina J, Narayanan T, Gao G, Rohde M, Tsentalovich D, Pasquali M, et al. Electrically insulating thermal nano-oils using 2D fillers. ACS Nano. 2012;**6**(2):1214-1220. DOI: 10.1021/nn203862p

[3] Hong Y, Lee E, Choi J, Oh S, Haam S, Huh Y, et al. Gold nanorod-mediated photothermal modulation for localized ablation of cancer cells. Journal of Nanomaterials. 2012;**2012**:1-7. DOI: 10.1155/2012/825060

[4] Bianco A, Kostarelos K, Prato M. Applications of carbon nanotubes in drug delivery. Current Opinion in Chemical Biology. 2005;**9**(6):674-679. DOI: 10.1016/j.cbpa.2005.10.005

[5] Ansorena MR, Marcovich NE, Pereda M. Food biopackaging based on chitosan. In: Torres Martínez L, Kharissova O, Kharisov B, editors. Handbook of Ecomaterials. Cham: Springer; 2017. pp. 1-27. DOI: 10.1007/978-3-319-48281-1_68-1

[6] Zhang Q, Wu S, Zhang L, Lu J, Verproot F, Liu Y, et al. Fabrication of polymeric ionic liquid/graphene nanocomposite for glucose oxidase immobilization and direct electrochemistry. Biosensors and Bioelectronics. 2011;**26**(5):2632-2637. DOI: 10.1016/j.bios.2010.11.024

[7] Shao Y, Wang J, Wu H, Liu J, Aksay I, Lin Y. Graphene based electrochemical sensors and biosensors: A Review. Electroanalysis. 2010;**22**(10):1027-1036. DOI: 10.1002/elan.200900571

[8] Dong H, Ding L, Yan F, Ji H, Ju H. The use of polyethylenimine-grafted graphene nanoribbon for cellular delivery of locked nucleic acid modified molecular beacon for recognition of microRNA. Biomaterials. 2011;**32**(15):3875-3882. DOI: 10.1016/j.biomaterials.2011.02.001

[9] Zhu M, Wang R, Nie G. Applications of nanomaterials as vaccine adjuvants. Human Vaccines & Immunotherapeutics. 2014;**10**(9):2761-2774. DOI: 10.4161/hv.29589

[10] Ciofani G, Raffa V, Menciassi A, Dario P. Preparation of boron nitride nanotubes aqueous dispersions for biological pplications. Journal of Nanoscience and Nanotechnology. 2008;**8**(12):6223-6231. DOI: 10.1166/jnn.2008.339

[11] Chen X, Wu P, Rousseas M, Okawa D, Gartner Z, Zettl A, et al. Boron nitride nanotubes are noncytotoxic and can be functionalized for interaction with proteins and cells. Journal of the American Chemical Society. 2009;**131**(3):890-891. DOI: 10.1021/ja807334b

[12] Ribeiro H, Trigueiro J, Silva W, Woellner C, Owuor P, Cristian Chipara A, et al. Hybrid MoS2/h-BN nanofillers as synergic heat dissipation and reinforcement additives in epoxy nanocomposites. ACS Applied Materials & Interfaces. 2017. DOI: 10.1021/acsami.7b09945. https://pubs.acs.org/doi/10.1021/acsami.7b09945

[13] Ribeiro H, Trigueiro J, Lopes M, Pedrotti J, Woellner C, Silva W, et al. Enhanced thermal conductivity and mechanical properties of hybrid MoS2/h-BN polyurethane nanocomposites. Journal of Applied Polymer Science. 2018;**135**(30):46560. DOI: 10.1002/app.46560

[14] Ribeiro H, Trigueiro J, Owuor P, Machado L, Woellner C, Pedrotti J, et al. Hybrid 2D nanostructures for

mechanical reinforcement and thermal conductivity enhancement in polymer composites. Composites Science and Technology. 2018;**159**:103-110. DOI: 10.1016/j.compscitech.2018.01.032

[15] Emanet M, Şen Ö, Çulha M. Evaluation of boron nitride nanotubes and hexagonal boron nitrides as nanocarriers for cancer drugs. Nanomedicine. 2017;**12**(7):797-810. DOI: 10.2217/nnm-2016-0322

[16] Kuila T, Bose S, Mishra A, Khanra P, Kim N, Lee J. Chemical functionalization of graphene and its applications. Progress in Materials Science. 2012;**57**(7):1061-1105. DOI: 10.1016/j.pmatsci.2012.03.002

[17] Tang P, Hu G, Gao Y, Li W, Yao S, Liu Z, et al. The microwave adsorption behavior and microwave-assisted heteroatoms doping of graphene-based nano-carbon materials. Scientific Reports. 2014;**4**(1). DOI: 10.1038/srep05901

[18] Ribeiro H, da Silva W, Neves J, Calado H, Paniago R, Seara L, et al. Multifunctional nanocomposites based on tetraethylenepentamine-modified graphene oxide/epoxy. Polymer Testing. 2015;**43**:182-192. DOI: 10.1016/j.polymertesting.2015.03.010

[19] Sharma P, Tuteja S, Bhalla V, Shekhawat G, Dravid V, Suri C. Bio-functionalized graphene–graphene oxide nanocomposite based electrochemical immunosensing. Biosensors and Bioelectronics. 2013;**39**(1):99-105. DOI: 10.1016/j.bios.2012.06.061

[20] Kuila T, Khanra P, Mishra A, Kim N, Lee J. Functionalized-graphene/ethylene vinyl acetate co-polymer composites for improved mechanical and thermal properties. Polymer Testing. 2012;**31**(2):282-289. DOI: 10.1016/j.polymertesting.2011.12.003

[21] Valentini L, Cardinali M, Fortunati E, Torre L, Kenny J. A novel method to prepare conductive nanocrystalline cellulose/graphene oxide composite films. Materials Letters. 2013;**105**:4-7. DOI: 10.1016/j.matlet.2013.04.034

[22] Gao K, Shao Z, Wu X, Wang X, Li J, Zhang Y, et al. Cellulose nanofibers/reduced graphene oxide flexible transparent conductive paper. Carbohydrate Polymers. 2013;**97**(1):243-251. DOI: 10.1016/j.carbpol.2013.03.067

[23] Ke G, Guan W, Tang C, Guan W, Zeng D, Deng F. Covalent functionalization of multiwalled carbon nanotubes with a low molecular weight chitosan. Biomacromolecules. 2007;**8**(2):322-326. DOI: 10.1021/bm0604146

[24] Jiang Y, Zhang Q, Li F, Niu L. Glucose oxidase and graphene bionanocomposite bridged by ionic liquid unit for glucose biosensing application. Sensors and Actuators B: Chemical. 2012;**161**(1):728-733. DOI: 10.1016/j.snb.2011.11.023

[25] Liu Z, Robinson J, Sun X, Dai H. PEGylated nanographene oxide for delivery of water-insoluble cancer drugs. Journal of the American Chemical Society. 2008;**130**(33):10876-10877. DOI: 10.1021/ja803688x

[26] Zheng M, Jagota A, Semke E, Diner B, Mclean R, Lustig S, et al. DNA-assisted dispersion and separation of carbon nanotubes. Nature Materials. 2003;**2**(5):338-342. DOI: 10.1038/nmat877

[27] Ciofani G, Raffa V, Menciassi A, Dario P. Preparation of boron nitride nanotubes aqueous dispersions for biological applications. Journal of Nanoscience and Nanotechnology. 2008;**8**(12):6223-6231. DOI: 10.1166/jnn.2008.339

[28] Ciofani G, Danti S, Genchi G, Mazzolai B, Mattoli V. Boron nitride nanotubes: Biocompatibility and

potential spill-over in nanomedicine. Small. 2013;**9**(9-10):1672-1685. DOI: 10.1002/smll.201201315

[29] Ferreira T, Silva P, Santos R, Sousa E. A novel synthesis route to produce boron nitride nanotubes for bioapplications. Journal of Biomaterials and Nanobiotechnology. 2011;**02**(04):426-434. DOI: 10.4236/jbnb.2011.24052

[30] da Silva W, Hilário Ferreira T, de Morais C, Soares Leal A, Barros Sousa E. Samarium doped boron nitride nanotubes. Applied Radiation and Isotopes. 2018;**131**:30-35. DOI: 10.1016/j.apradiso.2017.10.045

[31] Dorato M, Buckley L. Toxicology testing in drug discovery and development. Current Protocols in Toxicology. 2007. DOI: 10.1002/0471141755.tx1901s31

[32] Fellows M, O'Donovan M. Cytotoxicity in cultured mammalian cells is a function of the method used to estimate it. Mutagenesis. 2007;**22**(4):275-280. DOI: 10.1093/mutage/gem013

[33] Rocca A, Marino A, Del Turco S, Cappello V, Parlanti P, Pellegrino M, et al. Pectin-coated boron nitride nanotubes: In vitro cyto−/immune-compatibility on RAW 264.7 macrophages. Biochimica et Biophysica Acta. 2016;**1860**(4):775-784. DOI: 10.1016/j.bbagen.2016.01.020. General Subjects

[34] Horváth L, Magrez A, Golberg D, Zhi C, Bando Y, Smajda R, et al. In vitro investigation of the cellular toxicity of boron nitride nanotubes. ACS Nano. 2011;**5**(5):3800-3810. DOI: 10.1021/nn200139h

[35] Del Turco S, Ciofani G, Cappello V, Gemmi M, Cervelli T, Saponaro C, et al. Cytocompatibility evaluation of glycol-chitosan coated boron nitride nanotubes in human endothelial cells. Colloids and Surfaces B: Biointerfaces. 2013;**111**:142-149. DOI: 10.1016/j.colsurfb.2013.05.031

[36] Hanagata N, Zhang H, Chen S, Zhi C, Yamazaki T. Chitosan-coated boron nitride nanospheres enhance delivery of CpG oligodeoxynucleotides and induction of cytokines. International Journal of Nanomedicine. 2013:1783. DOI: 10.2147/ijn.s43251

[37] Ciofani G, Raffa V, Menciassi A, Cuschieri A. Cytocompatibility, interactions, and uptake of polyethyleneimine-coated boron nitride nanotubes by living cells: Confirmation of their potential for biomedical applications. Biotechnology and Bioengineering. 2008;**101**(4):850-858. DOI: 10.1002/bit.21952

[38] Kıvanç M, Barutca B, Koparal A, Göncü Y, Bostancı S, Ay N. Effects of hexagonal boron nitride nanoparticles on antimicrobial and antibiofilm activities, cell viability. Materials Science and Engineering C. 2018;**91**:115-124. DOI: 10.1016/j.msec.2018.05.028

[39] Ciofani G, Danti S, Nitti S, Mazzolai B, Mattoli V, Giorgi M. Biocompatibility of boron nitride nanotubes: An up-date of in vivo toxicological investigation. International Journal of Pharmaceutics. 2013;**444**(1-2):85-88. DOI: 10.1016/j.ijpharm.2013.01.037

[40] Salvetti A, Rossi L, Iacopetti P, Li X, Nitti S, Pellegrino T, et al. In vivo biocompatibility of boron nitride nanotubes: Effects on stem cell biology and tissue regeneration in planarians. Nanomedicine. 2015;**10**(12):1911-1922. DOI: 10.2217/nnm.15.46

[41] Ferreira T, Hollanda L, Lancellotti M, de Sousa E. Boron nitride nanotubes chemically functionalized with glycol chitosan for gene transfection in eukaryotic cell lines. Journal of Biomedical Materials Research Part A.

2014;**103**(6):2176-2185. DOI: 10.1002/jbm.a.35333

[42] Emanet M, Kazanç E, Çobandede Z, Çulha M. Boron nitride nanotubes enhance properties of chitosan-based scaffolds. Carbohydrate Polymers. 2016;**151**:313-320. DOI: 10.1016/j.carbpol.2016.05.074

[43] Shuai C, Han Z, Feng P, Gao C, Xiao T, Peng S. Akermanite scaffolds reinforced with boron nitride nanosheets in bone tissue engineering. Journal of Materials Science: Materials in Medicine. 2015;**26**(5). DOI: 10.1007/s10856-015-5513-4

[44] Matson M, Wilson L. Nanotechnology and MRI contrast enhancement. Future Medicinal Chemistry. 2010;**2**(3):491-502. DOI: 10.4155/fmc.10.3

[45] Menichetti L, De Marchi D, Calucci L, Ciofani G, Menciassi A, Forte C. Boron nitride nanotubes for boron neutron capture therapy as contrast agents in magnetic resonance imaging at 3T. Applied Radiation and Isotopes. 2011;**69**(12):1725-1727. DOI: 10.1016/j.apradiso.2011.02.032

[46] Feofanova N, Geraldo J, Andrade L. Radiation oncology in vitro: Trends to improve radiotherapy through molecular targets. BioMed Research International. 2014;**2014**:1-13. DOI: 10.1155/2014/461687

[47] Perez C, Brady L. Principles and practice of radiation oncology. Journal of Pediatric Hematology/Oncology. 1999;**21**(6):560. DOI: 10.1097/00043426-199911000-00025

[48] Ferreira T, Marino A, Rocca A, Liakos I, Nitti S, Athanassiou A, et al. Folate-grafted boron nitride nanotubes: Possible exploitation in cancer therapy. International Journal of Pharmaceutics. 2015;**481**(1-2):56-63. DOI: 10.1016/j.ijpharm.2015.01.048

[49] Ciofani G, Raffa V, Menciassi A, Cuschieri A. Folate functionalized boron nitride nanotubes and their selective uptake by glioblastoma multiforme cells: Implications for their use as boron carriers in clinical boron neutron capture therapy. Nanoscale Research Letters. 2008;**4**(2):113-121. DOI: 10.1007/s11671-008-9210-9

[50] Nakamura H, Koganei H, Miyoshi T, Sakurai Y, Ono K, Suzuki M. Antitumor effect of boron nitride nanotubes in combination with thermal neutron irradiation on BNCT. Bioorganic & Medicinal Chemistry Letters. 2015;**25**(2):172-174. DOI: 10.1016/j.bmcl.2014.12.005

[51] Ciani L, Bortolussi S, Postuma I, Cansolino L, Ferrari C, Panza L, et al. Rational design of gold nanoparticles functionalized with carboranes for application in boron neutron capture therapy. International Journal of Pharmaceutics. 2013;**458**(2):340-346. DOI: 10.1016/j.ijpharm.2013.10.008

[52] Asano R, Nagami A, Fukumoto Y, Miura K, Yazama F, Ito H, et al. Synthesis and biological evaluation of new boron-containing chlorin derivatives as agents for both photodynamic therapy and boron neutron capture therapy of cancer. Bioorganic & Medicinal Chemistry Letters. 2014;**24**(5):1339-1343. DOI: 10.1016/j.bmcl.2014.01.054

[53] Li X, Hanagata N, Wang X, Yamaguchi M, Yi W, Bando Y, et al. Multimodal luminescent-magnetic boron nitride nanotubes@NaGdF4:Eu structures for cancer therapy. Chemical Communications. 2014;**50**(33):4371-4374. DOI: 10.1039/c4cc00990h

[54] Li X, Wang X, Zhang J, Hanagata N, Wang X, Weng Q, et al. Hollow boron nitride nanospheres as boron reservoir for prostate cancer treatment. Nature

Communications. 2017;**8**:13936. DOI: 10.1038/ncomms13936

[55] Wang J, Lee C, Yap Y. Recent advancements in boron nitride nanotubes. Nanoscale. 2010;**2**(10):2028. DOI: 10.1039/c0nr00335b

[56] Ciofani G, Raffa V, Menciassi A, Cuschieri A. Boron nitride nanotubes: An innovative tool for nanomedicine. Nano Today. 2009;**4**(1):8-10. DOI: 10.1016/j.nantod.2008.09.001

[57] Raffa V, Ciofani G, Cuschieri A. Enhanced low voltage cell electropermeabilization by boron nitride nanotubes. Nanotechnology. 2009;**20**(7):075104. DOI: 10.1088/0957-4484/20/7/075104

[58] Versiani A, Andrade L, Martins E, Scalzo S, Geraldo J, Chaves C, et al. Gold nanoparticles and their applications in biomedicine. Future Virology. 2016;**11**(4):293-309. DOI: 10.2217/fvl-2015-0010

[59] Fiorillo M, Verre AF, Iliut M, Peiris-Pagés M, Ozsvari B, Gandara R, et al. Graphene oxide selectively targets cancer stem cells, across multiple tumor types: Implications for non-toxic cancer treatment, via "differentiation-based nano-therapy". Oncotarget. 2015;**6**(6). DOI: 10.18632/oncotarget.3348

[60] Gray A, Harlen O, Harris S, Khalid S, Leung Y, Lonsdale R, et al. In pursuit of an accurate spatial and temporal model of biomolecules at the atomistic level: a perspective on computer simulation. Acta Crystallographica, Section D: Biological Crystallography. 2015;**71**(1):162-172. DOI: 10.1107/s1399004714026777

[61] Proctor E, Dokholyan N. Applications of discrete molecular dynamics in biology and medicine. Current Opinion in Structural Biology. 2016;**37**:9-13. DOI: 10.1016/j.sbi.2015.11.001

[62] Vaidehi N, Jain A. Internal coordinate molecular dynamics: A foundation for multiscale dynamics. Journal of Physical Chemistry B. 2015;**119**(4):1233-1242. DOI: 10.1021/jp509136y

[63] Perilla J, Goh B, Cassidy C, Liu B, Bernardi R, Rudack T, et al. Molecular dynamics simulations of large macromolecular complexes. Current Opinion in Structural Biology. 2015;**31**:64-74. DOI: 10.1016/j.sbi.2015.03.007

[64] Meneksedag-Erol D, Tang T, Uludağ H. Molecular modeling of polynucleotide complexes. Biomaterials. 2014;**35**(25):7068-7076. DOI: 10.1016/j.biomaterials.2014.04.103

[65] Bergonzo C, Galindo-Murillo R, Cheatham T. Molecular modeling of nucleic acid structure: energy and sampling. Current Protocols in Nucleic Acid Chemistry. 2013:7.8.1-7.8.21. DOI: 10.1002/0471142700.nc0708s54

[66] van Mourik T, Buhl M, Gaigeot M. Density functional theory across chemistry, physics and biology. Philosophical Transactions of the Royal Society A: Mathematical, Physical and Engineering Sciences. 2014;**372**(2011):20120488-20120488. DOI: 10.1098/rsta.2012.0488

[67] Zhang L, Wang X. DNA sequencing by hexagonal boron nitride nanopore: A computational study. Nanomaterials. 2016;**6**(6):111. DOI: 10.3390/nano6060111

[68] Zhou Z, Hu Y, Wang H, Xu Z, Wang W, Bai X, et al. DNA translocation through hydrophilic nanopore in hexagonal boron nitride. Scientific Reports. 2013;**3**(1). DOI: 10.1038/srep03287

[69] Briggs K, Madejski G, Magill M, Kastritis K, de Haan H, McGrath J, et al.

DNA translocations through nanopores under nanoscale preconfinement. Nano Letters. 2017;**18**(2):660-668. DOI: 10.1021/acs.nanolett.7b03987

[70] Liu S, Lu B, Zhao Q, Li J, Gao T, Chen Y, et al. Boron nitride nanopores: Highly sensitive DNA single-molecule detectors. Advanced Materials. 2013;**25**(33):4549-4554. DOI: 10.1002/adma.201301336

[71] Liang L, Li J, Zhang L, Zhang Z, Shen J, Li L, et al. Computer simulation of water desalination through boron nitride nanotubes. Physical Chemistry Chemical Physics. 2017;**19**(44):30031-30038. DOI: 10.1039/c7cp06230c

[72] Azamat J, Khataee A. Molecular dynamics simulations of removal of cyanide from aqueous solution using boron nitride nanotubes. Computational Materials Science. 2017;**128**:8-14. DOI: 10.1016/j.commatsci.2016.10.040

[73] Azamat J, Sattary B, Khataee A, Joo S. Removal of a hazardous heavy metal from aqueous solution using functionalized graphene and boron nitride nanosheets: Insights from simulations. Journal of Molecular Graphics and Modelling. 2015;**61**:13-20. DOI: 10.1016/j.jmgm.2015.06.012

[74] Azamat J, Khataee A, Joo S. Molecular dynamics simulations of trihalomethanes removal from water using boron nitride nanosheets. Journal of Molecular Modeling. 2016;**22**(4). DOI: 10.1007/s00894-016-2939-7

[75] Tolladay M, Ivanov D, Allan N, Scarpa F. Piezoelectric effects in boron nitride nanotubes predicted by the atomistic finite element method and molecular mechanics. Nanotechnology. 2017;**28**(35):355705. DOI: 10.1088/1361-6528/aa765b

[76] Falin A, Cai Q, Santos E, Scullion D, Qian D, Zhang R, et al. Mechanical properties of atomically thin boron nitride and the role of interlayer interactions. Nature Communications. 2017;**8**:15815. DOI: 10.1038/ncomms15815

[77] Roosta S, Hashemianzadeh S, Ketabi S. Encapsulation of cisplatin as an anti-cancer drug into boron-nitride and carbon nanotubes: Molecular simulation and free energy calculation. Materials Science and Engineering: C. 2016;**67**:98-103. DOI: 10.1016/j.msec.2016.04.100

[78] Roosta S, Nikkhah S, Sabzali M, Hashemianzadeh S. Molecular dynamics simulation study of boron-nitride nanotubes as a drug carrier: From encapsulation to releasing. RSC Advances. 2016;**6**(11):9344-9351. DOI: 10.1039/c5ra22945f

[79] Mehrjouei E, Akbarzadeh H, Shamkhali A, Abbaspour M, Salemi S, Abdi P. Delivery of cisplatin anti-cancer drug from carbon, boron nitride, and silicon carbide nanotubes forced by Ag-nanowire: A comprehensive molecular dynamics study. Molecular Pharmaceutics. 2017;**14**(7):2273-2284. DOI: 10.1021/acs.molpharmaceut.7b00106

[80] Acconcia F, Pallottini V, Marino M. Molecular mechanisms of action of BPA. Dose-Response. 2015;**13**(4):155932581561058. DOI: 10.1177/1559325815610582

[81] Tahover E, Patil Y, Gabizon A. Emerging delivery systems to reduce doxorubicin cardiotoxicity and improve therapeutic index. Anti-Cancer Drugs. 2015;**26**(3):241-258. DOI: 10.1097/cad.0000000000000182

[82] Zeng Z, Zhao P, Liu L, Gao X, Mao H, Chen Y. Lipid stabilized solid drug nanoparticles for targeted chemotherapy. ACS Applied Materials & Interfaces. 2018;**10**(30):24969-24974. DOI: 10.1021/acsami.8b07024

[83] Khatti Z, Hashemianzadeh S. Boron nitride nanotube as a delivery system

for platinum drugs: Drug encapsulation and diffusion coefficient prediction. European Journal of Pharmaceutical Sciences. 2016;**88**:291-297. DOI: 10.1016/j.ejps.2016.04.011

[84] Duverger E, Gharbi T, Delabrousse E, Picaud F. Quantum study of boron nitride nanotubes functionalized with anticancer molecules. Physical Chemistry Chemical Physics. 2014;**16**(34):18425-18432. DOI: 10.1039/c4cp01660b

[85] El Khalifi M, Bentin J, Duverger E, Gharbi T, Boulahdour H, Picaud F. Encapsulation capacity and natural payload delivery of an anticancer drug from boron nitride nanotube. Physical Chemistry Chemical Physics. 2016;**18**(36):24994-25001. DOI: 10.1039/c6cp01387b

Chapter 3

Bio-Inspired Artificial Light-Harvesting Arrays Based on Boron(III)-Chelates

Hatun H.T. Al-Sharif and Anthony Harriman

Abstract

Many diverse natural systems use sunlight to drive critical chemical reactions. To harvest sufficient photon densities, natural organisms have developed highly sophisticated light absorbing antennae rather than rely on direct illumination of a single chromophore. Attempts to develop artificial analogues have resulted in the synthesis and spectroscopic characterisation of elaborate molecular assemblies and here we consider the case for using boron(III) chelates as the primary light absorbers. Such entities make attractive modules for the creation of multi-component arrays with individual units sited in a logical sequence for long-range electronic energy transfer. Alternatively, certain boron(III) chelates can be synthesised in high yield by simple strategies that avoid time-consuming purification. These latter materials are appealing as components for large-scale light harvesters. The use of photonic crystals avoids the need to position individual molecules at the catalyst but presents severe design challenges. Interrupting, or redirecting, the flow of excitons within the array requires the introduction of novel switches that can be activated by selective illumination. Protecting the array against adventitious photofading is a major objective that has yet to be achieved.

Keywords: fluorescence, electronic energy transfer, photophysics, chromophores, covalent networks, photonic crystals, photosynthesis

1. Introduction

Most chemical reactions that occur in biology have to be performed under mild conditions. With few exceptions, this means running the reaction at atmospheric pressure and temperatures at around 37°C. The main solvent is water. Despite such severe restrictions, Nature is able to supply an inordinately wide variety of chemicals on an enormous scale. Extensive use is made of enzyme catalysts, templates and protein matrices to assemble the required materials and to ensure selectivity. A further challenge inherent for much of natural synthesis is the need to produce chiral products with high specificity. During evolution, Nature has learned to re-use a basic molecular structure for multiple purposes simply by making minor, but critical, changes. The other key trick used by Nature is to arrange a limited number of building blocks in different sequences so as to engineer unique structures comprising only a handful of simple modules. The most illustrious examples of this particular strategy are the polynucleotides.

IntechOpen

By way of slow evolution, Nature has learned many other tricks. One that is relevant here concerns the attachment of light-harvesting proteins to catalytic sites so as to achieve high rates of photochemical synthesis. Photochemistry is surely the cleanest, simplest and most effective means for driving high-energy reactions. Such processes invariably require a catalyst to perform bond making, or maybe bond breaking, steps. These enzymes are optimised for the difficult thermal chemistry, which might involve multi-electron steps, and usually lack the chromophores needed to absorb sunlight at a reasonable level. The answer to this conundrum is to append a light harvester to the enzyme with the sole purpose of channelling absorbed photons to the active site. This approach is employed throughout all biological processes that use light to initiate a chemical reaction. The light harvester ranges from a single molecule, as found in photolyase DNA repair enzymes, to the intricate molecular machinery developed for photosynthetic bacteria and plants. These latter systems involve several hundreds of individual chromophores arranged cooperatively to collect photons of disparate frequency and direct the exciton to the correct location.

Although the structure and function of many natural light-harvesting systems has been reviewed in detail, much remains to be learned about the assembly process. Some photon collectors are self-repairing and contain built-in redundancy to compensate for inevitable damage during operation. This is the case for the Fenna-Matthews-Olson complex found in photosynthetic bacteria [1]. The natural systems are often equipped with ancillary protective mechanisms. For example, light-harvesting membranes contain high levels of carotenoids [2] to protect against accidental generation of oxy-radicals and singlet molecular oxygen. The other critical issue to emerge from a study of the natural units is the realisation of the staggering amount of material involved. We can illustrate this point by considering the following features of the biological world:

- Wheatgrass contains up to 70% by mass of chlorophyll, the main pigment of natural photosynthesis;

- Green plants contain several different types of chlorophyll, each possessing a different absorption spectrum to maximise overall light collection;

- Photosynthetic bacteria contain roughly 200–250 molecules of chlorophyll for each catalytic site responsible for fuel production;

- Reliable methods for measuring chlorophyll content indicate levels ranging from 41 mg m^{-2} up to 675 mg m^{-2} according to geographical location and type of organism;

- The 45-acre site at Clayhill in Bedfordshire, one of the UK's operational solar farms, generates enough power for 2500 homes. Assuming 50% of the acreage is taken up with solar panels, this would be the equivalent of at least 10 kg of chlorophyll;

- The current cost of BODIPY (see below) is £170 per gram and a typical synthesis might produce 2 g of purified material.

The main conclusion drawn from this crude analysis is that protocols for the efficient synthesis of the dye are as important as the optical properties. Indeed, it is essential that high yields of compounds for artificial light-harvesting applications are attainable under conditions that do not require elaborate, or time-consuming, purification procedures. The basic operating principle for an artificial light-harvesting array is sketched in **Figure 1**. Individual chromophores are arranged in such a way that each compound can absorb a fraction of incident sunlight. To provide for panchromatic absorption, a minimum of three separate chromophores is required while each absorber needs to be equipped with ancillary chromophores to broaden the spectral response. The resultant exciton can be transferred around the network until reaching the terminal acceptor where it can be off-loaded to the device. The latter removes the exciton from the array so as to fulfil some useful task.

We now enquire about the possibility of constructing simple artificial light-harvesting systems using tetrahedral boron(III) chelates as the generic building block. These chromophores are based on small repeat units in which the boron atom plays the critical role of assembling the unit into a macrocyclic structure with extended π-conjugation running along the molecular backbone. A variety of such units is available. Each unit can be equipped with secondary features to aid solubility or modulate physiochemical properties. Units can be linked together to create new electronic entities with synergistic performance. Chirality can be introduced by several different means, if required. Below is given a generic model to better illustrate the design features (**Figure 2**). One such molecule acting alone is never going to solve the light-harvesting problems and so it is necessary to import some means for selective accretion of molecules into patterned arrays. At a rough guess, it might be necessary to equip each catalyst with 100 or more chromophores to ensure a high photon flux reaching the reactive site. This realisation adds further complexity to the final design, which has to go much further than simply identifying some useful building blocks. It is clear, even at the beginning of our search, that the individual units must be cheap, stable, safe-to-handle and easily recycled. A further requirement is the need to develop a repair strategy but this might be too advanced for the moment! We begin by considering likely boron(III) chelated modules.

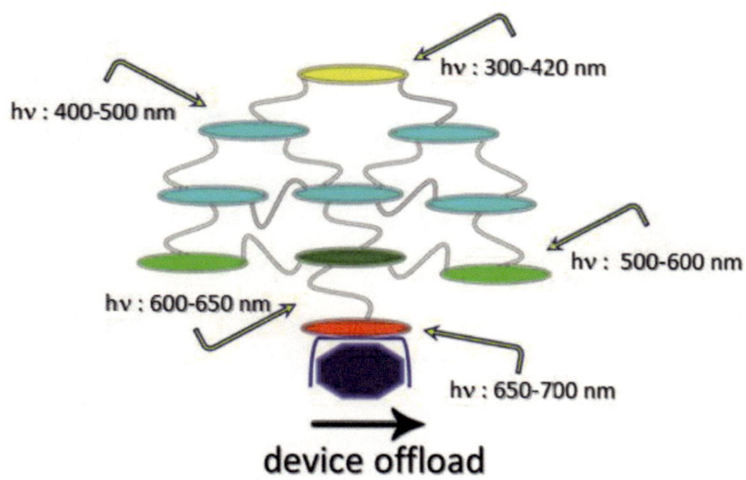

Figure 1.
Pictorial representation of a putative artificial light-harvesting antenna. Each disc is intended to represent a boron(III) chelate equipped with ancillary chromophores. Each of these discs is designed to collect photons over a particular wavelength range. Electronic energy transfer can occur between the various discs; this will be reversible for identical chromophores but will be irreversible when there is an energy gradient. The final boron(III) chelate (coloured red on the scheme) is closely associated with a device, which removes the photon from the array. Individual EET steps must be fast to avoid photofading of the chromophores. The varied routes available to EET reflect the need for redundancy.

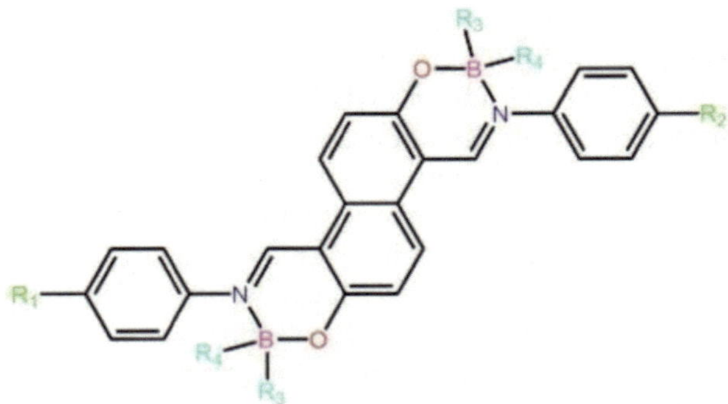

Figure 2.
Illustration of the functionalisation of the basic boron(III) chelate used as principle light absorber. The various sites indicated by the symbol "R" can be used to improve physiochemical properties, such as solubility, and to tune the optical properties. This level of derivatisation is necessary for identifying optimum photonic crystals.

2. Boron dipyrromethene (BODIPY) chromophores

Without doubt, the family of boron dipyrromethene (BODIPY) [3, 4] dyes has inherited the title of "Queen of Fluorophores" over the past decade or so. The basic structural framework facilitates almost limitless opportunities for modification and expansion, leading to a bewildering variety of compounds. The conventional structure, shown in **Figure 3**, has a planar geometry around the dipyrrin unit imposed by the tetrahedral boron(III) site. In general, two types of "conventional" BODIPY dyes are distinguishable according to the provision, or not, of an unconstrained *meso*-aryl ring. Steric clashes between adjacent hydrogen atoms restrict rotation of

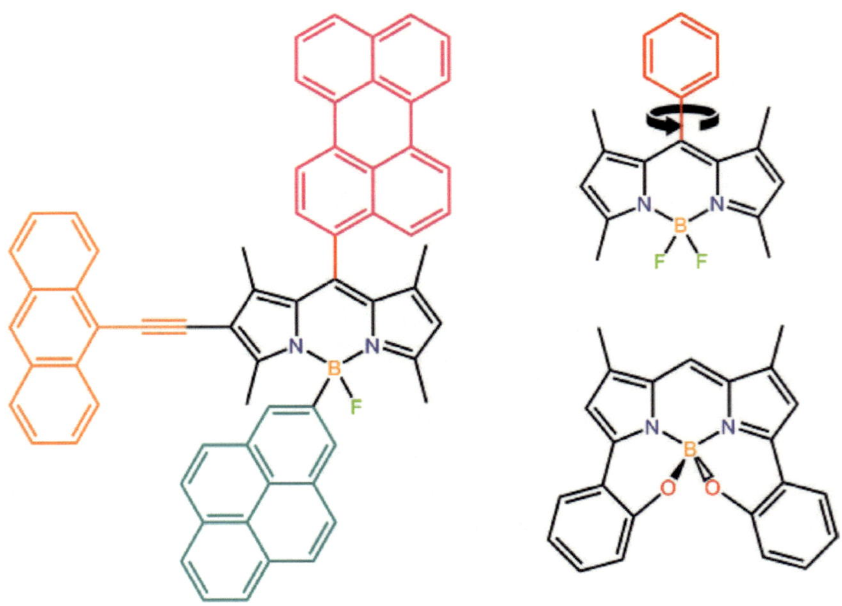

Figure 3.
The upper right-hand panel represent a conventional BODIPY dye equipped with a rigidifying BF_2 residue. Substituents can replace the hydrogen atoms at any of the carbon atoms while it is possible to exchange one or both fluorine groups with a wide variety of organic residues. The meso-site, otherwise known as the 8-position, is especially interesting in as much as aryl hydrocarbons can be attached and used as energy donors. The lower right-hand side shows a chiral derivative. The left-hand panel shows how the basic BODIPY nucleus can be equipped with ancillary light harvesters so as to provide for panchromatic photon collection.

this aryl ring around the connecting C-C bond and this leads to modest distortion of the dipyrrin backbone. The result is an interesting class of fluorophores known [5] as "molecular rotors" that can be used to report on rheology changes. These compounds tend not to be strongly fluorescent and are of limited value in terms of photon concentrators. The other type of conventional BODIPY fluorophore uses alkyl groups to block gyration of any *meso*-aryl ring and the resultant compounds tend to be highly emissive under all conditions. It is a simple matter to disperse these chromophores in plastic sheets to give highly effective luminescent solar concentrators, where a high fraction of the fluorescence appears at the edges of the film. Suitable attachment to a solar panel provides the means to sensitise the semiconductor towards particular segments of the solar spectrum. Such devices have particular relevance to organic solar cells.

These classical BODIPY dyes are robust, easily functionalised, photochemically stable and strongly fluorescent. Typical absorption and fluorescence spectral maxima lie in the 470–550 region. Fluorescence quantum yields tend towards unity while excited-state lifetimes are typically around 5 ns. Lindsey et al. reported the first example of the incorporation of such BODIPY dyes into artificial light-harvesting systems in 1994 [6, 7]. Their system used a terminal BODIPY dye to inject excitons into a linear array of porphyrinic chromophores (**Figure 4**). Electronic energy migration and transfer results in long-range transport of the exciton from the BODIPY dye to the distal terminal, supplemented by secondary light absorption by the zinc porphyrins. This system was the starting point for an avalanche of related arrays capable of vectorial exciton migration. The main design principle for these systems is to create a gradient of excitation energies that facilitates stepwise transfer of the exciton along a molecular-scale wire.

Since the original report, a wealth of such artificial arrays has appeared in the literature, synthesised by many different research groups. The common goal is to use a series of disparate chromophores to absorb different regions of the solar spectrum and transfer the excitation energy along a linear or branched chain. A certain fraction of the energy is lost at each stage but, by careful positioning of the reagents, very fast (i.e., sub-ns) electronic energy transfer can be engineered. By judicious choice of chromophore, energy transfer can be unidirectional and almost quantitative [7]. The exciton ultimately reaches a terminal site where it needs to be off-loaded to some kind of device, such as a solar cell or light-active catalyst (**Figure 1**). Some of these arrays possess elaborate three-dimensional structures that can collect considerable numbers of photons over a wavelength range spanning many hundreds of nanometres. One of the largest such architectures is depicted in **Figure 5** and contains a total of 21 chromophores packed into a small volume. An added attraction of these artificial arrays is that the individual energy transfer steps can be monitored by time-resolved optical spectroscopy.

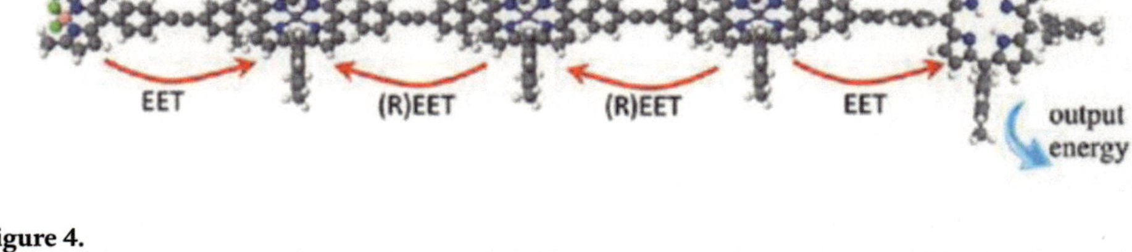

Figure 4.
Prototypic molecular-scale wire, comprising a BODIPY-based dye as input unit, three zinc porphyrins capable of forward and reverse energy transfer and a terminal free-base porphyrin as the emitting output unit. Reproduced from Ref. [7].

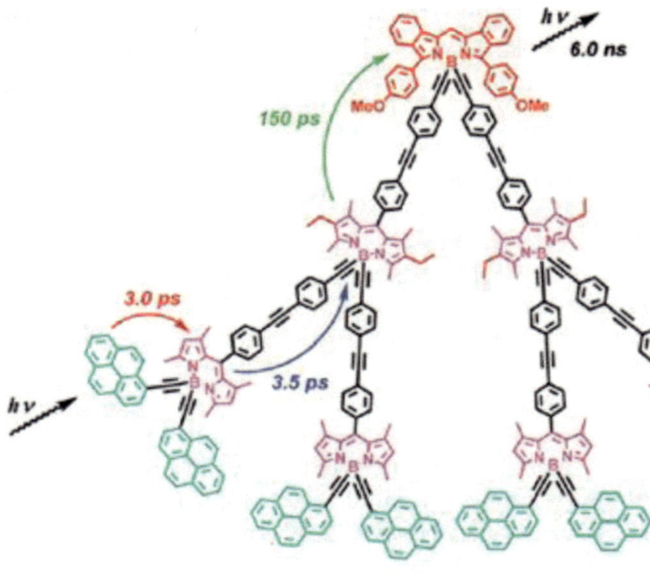

Figure 5.
Molecular formula for the large artificial light-harvesting array comprising 21 discrete chromophores, with the lifetimes indicated for each of the main units. Reproduced from Ref. [7].

It will not have escaped attention that these artificial light-harvesting arrays require quite sophisticated synthesis, involving multiple steps with low overall yields. They are expensive to produce and it seems unlikely that large quantities could be prepared by robotic methods. The latter is essential to generate sufficient quantities of material for practical application. Fortunately, the arrays seem to be relatively photostable. This is because the excited-state lifetimes of the dyes that contribute to the energy-transfer cascade are kept very short and also because the BODIPY nucleus does not favour intersystem crossing to the triplet manifold with reasonable efficacy. Only the terminal dye is susceptible to photobleaching and only if the deactivation of this unit is slow – for example, when the device is inoperable. This situation has been described by Stachelek et al. using a multi-component BODIPY-based array [8] (**Figure 6**). Here, the molecule undergoes stepwise photobleaching leading to successive loss of individual chromophores but retains photoactivity at all stages. The net result is that the array can absorb more than 10 million photons and still remain operational. Nonetheless, the protracted synthesis appears to rule out such systems as being viable for largescale artificial light-harvesters.

Lindsey et al. have questioned [9] if such arrays are effective as photon collectors. Their approach has involved detailed consideration of the required density of chromophores at a planar semi-conductor surface needed to effect a useful level of sensitization. This density has to take into account the absorptivity of the chromophore, its molecular volume, surface coverage and ancillary chromophores able to transfer excitation energy to the ultimate donor. Their general conclusion is not too encouraging for the practical application of covalently linked chromophoric arrays. This realisation has led the Lindsey Group to look at bio-hybrid systems whereby artificial pigments replace natural analogues in protein-based matrices. The intention is to make use of the protein to achieve very high chromophore densities having the minimum wastage of molecular volume.

Other research groups have sought alternatives to the covalent synthesis needed to assemble molecular architectures such as that shown in **Figure 4**. For example, supramolecular systems can be built by intercalating suitable chromophores into DNA. This avoids the need to link together the individual reagents via covalent

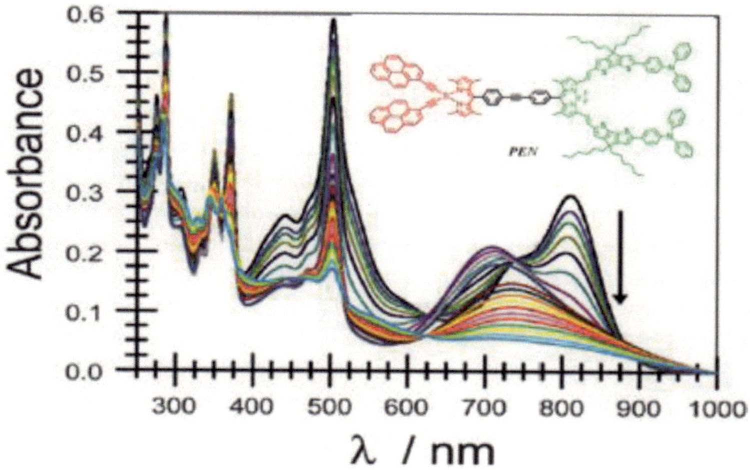

Figure 6.
Effect of continuous illumination with white light on the absorption spectrum recorded for PEN in deaerated CH_2Cl_2. Spectra were recorded at regular intervals over a period of 100 hours. Reproduced from Ref. [8].

synthesis but it becomes more challenging to position the reagents in a logical sequence. Other approaches to the same end include accretion of individual chromophores into aggregates that retain photoactivity. Such assemblies might be stacked columns of planar dyes or J-aggregates. Their main advantage is that many hundreds of chromophores can be self-assembled into one giant structure, driven simply by solvatophobic effects. A further benefit of this approach is that the columns can be easily dismantled and re-assembled in the event that an individual component becomes damaged during operation. This kind of replacement would be difficult, if not impossible, for the elaborate covalent networks.

3. Symmetrical pyrrole-BF_2 (BOPHY) modules

Over the past few years or so, the supremacy of BODIPY as a fluorescent reagent has been challenged by a series of new fluorophores, one of which is termed BOPHY [10] (**Figure 7**). This latter compound is easily synthesised in high yield and provides for formation of a wide variety of derivatives. It is not as versatile as BODIPY but is a very useful building block for preparation of linear molecules. Despite its longer π-conjugation pathway, the lowest-energy absorption maximum observed for BOPHY is blue shifted relative to that of BODIPY and is more complex. Unlike

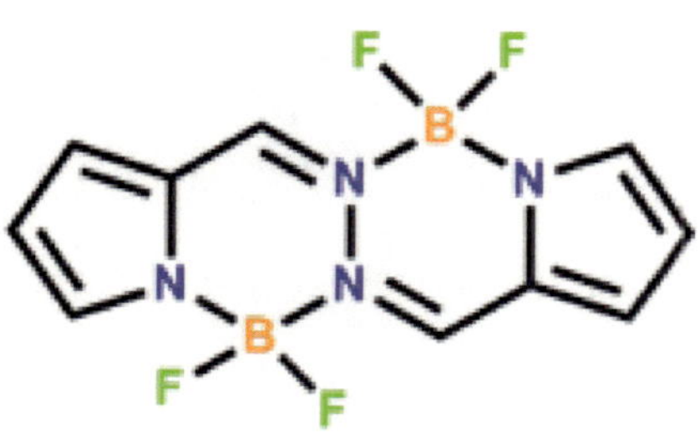

Figure 7.
The molecular formula for the basic BOPHY core, where different substituents can be appended at the various carbon atoms. Also shown is a solution of BOPHY in $CHCl_3$ solution with and without UV illumination. The final panel illustrates a fluorescent chemical sensor developed for the detection of copper(II) cations in fluid solution.

BODIPY, there is poor mirror symmetry between absorption and fluorescence spectral profiles, at least in the liquid phase [11]. However, the lowest-energy absorption transition is broadened relative to BODIPY and this is a useful feature for a putative photon collector. Recently, BOPHY has been used as the emissive segment of certain types of fluorescent sensor. The tetrahedral geometry around the BF_2 groups helps to minimise self-association and BOPHY-based compounds tend to be highly soluble in common organic solvents. Other studies have shown that bromination of BOPHY results in a compound able to sensitise formation of singlet molecular oxygen under illumination in fluid solution. Additional work has described the photochemical bleaching of BOPHY in liquid and solid phases. It appears that the compound is highly resistant to photofading under visible light illumination [12].

Ziessel et al. have reported [13] on the properties of the first molecular dyad built around the BOPHY scaffold (**Figure 8**). Here, a perylene moiety has been linked to the BOPHY core by way of an ethynylene group, thereby ensuring both close proximity and strong π-conjugation. These authors reported that direct excitation into the perylene group causes rapid electronic energy transfer to the BOPHY unit. The latter emits at long wavelength because of the increased conjugation. This raises the question as to whether exciton passage from perylene to BOPHY is internal conversion or intramolecular electronic energy transfer. Other fused ring BOPHY derivatives have appeared very recently where the main absorption transition can be tuned across the visible spectral range. These materials are useful building blocks for creation of multi-component light-harvesting arrays but such molecular assemblies have not yet been reported.

Related research has considered ways to design redox-active BOPHY derivatives [14]. The basic BOPHY framework is equipped with styryl units that extend the π-conjugation pathway in a linear fashion. The styryl units are terminated with N,N-dimethylamino groups that can be protonated or oxidised. This simple strategy allows the evaluation of long-range electronic interactions between the terminal groups as the charge increases. It might be stressed that adding conjugated groups to the BOPHY core is not too demanding in terms of synthesis and characterisation. However, there remains the problem of scale and questions about how best to produce large quantities of such material.

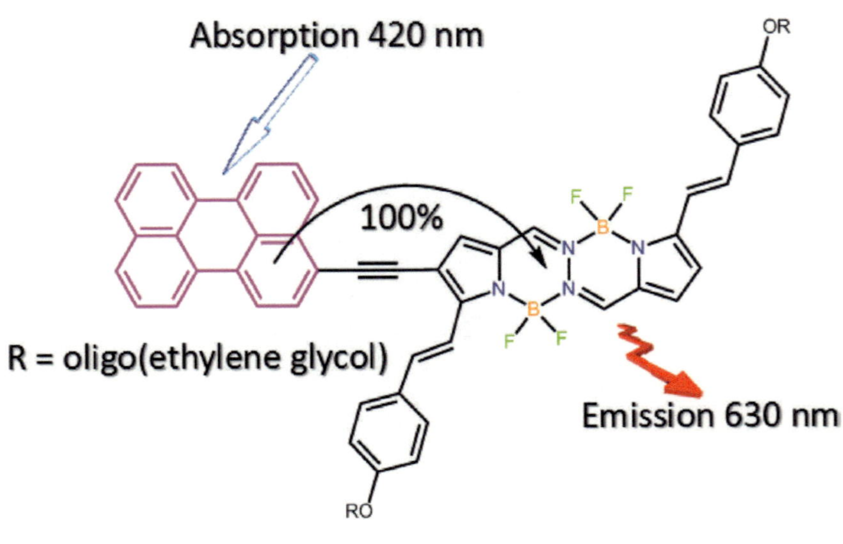

Figure 8.
Example of an artificial light-harvesting analogue constructed around the BOPHY chromophore. The light collection facility is expanded by adding a perylene function to the BOPHY unit. Excitons generated at the aryl hydrocarbon are transferred rapidly to the BOPHY unit, which itself emits at long wavelength.

One of the key features of the BOPHY family relates to its high fluorescence yield in the absence of perturbing groups. Fluorescence is almost quantitative for the simpler derivatives and BOPHY-based dyes are ideal candidates for inclusion within plastic luminescent solar concentrators. The S_0-S_1 absorption transition is primarily HOMO-LUMO in nature and the radiative rate constant is high. Triplet state formation is negligible for simple BOPHY derivatives in solution but certain substituents are able to promote intersystem crossing. A report has appeared that describes emission from certain crystals formed from thiophene-substituted BOPHY [15]. It is anticipated that many more functionalised BOPHY derivatives will appear in the near future.

4. Related boron(III)-chelated chromophores

Four-coordinate organo-boron complexes have emerged as promising materials for optoelectronics, such as organic light-emitting diodes (OLEDs) and organic field-effect transistors. This work is still at the exploratory stage but the new boron(III) chelates offer real prospects for the development of a range of blue emitters. This is a wavelength range where there are relatively few efficient fluorophores having the necessary stability and electroluminescence performance. This is especially true for deep blue emitters, in marked contrast to the numerous red and green emitters that are already in the marketplace. Since, in general, certain boranils and boron β-diketonates are known to exhibit high fluorescence quantum yields in the solid state, together with good levels of photostability, these materials are beginning to attract attention. Examples of such compounds include the *mono*-boranils shown below in **Figure 9**.

Ziessel, in particular, has conducted a detailed search for alternative fluorescent dyes to the popular BODIPY family [16]. Both the type of ligand and the nature of the assembling boron(III) fragment has been considered. This has led to several classes of tetrahedral boron(III) compounds as illustrated via **Figure 10**. The range of available compounds can be extended by considering five-membered rings with quinolines or six-membered rings with salicylaldehydes, oxazolylphenolates, acylpyrrole, or pyridinephenolates. Additional six-membered rings have been formed from chelation to phenalene-1,3-dione and perylenediimides. Many of these novel systems are easily adapted to form binuclear complexes, such as those prepared around naphthyl ligands (**Figure 11**). In all of these cases, the synthetic strategy used to isolate the compounds is straightforward and avoids the use of complicated procedures. Quite pure materials emerge from the crude reaction mixtures without the need for protracted column chromatography. Unfortunately,

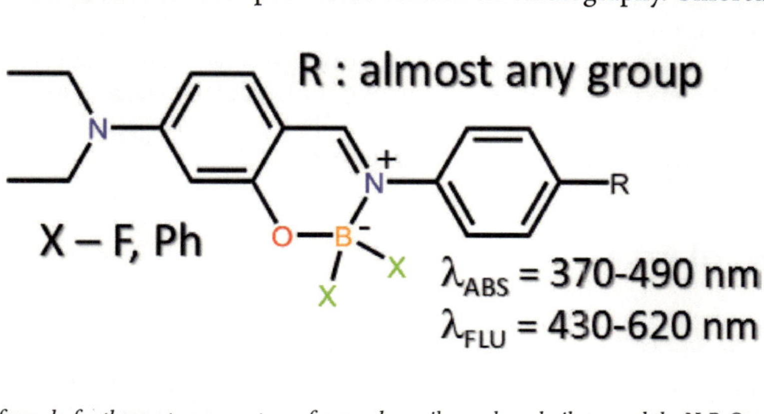

Figure 9.
Molecular formula for the most common type of mono-boranil complexes built around the N-B-O pattern. The terminal R group can be varied to include many different functionalities.

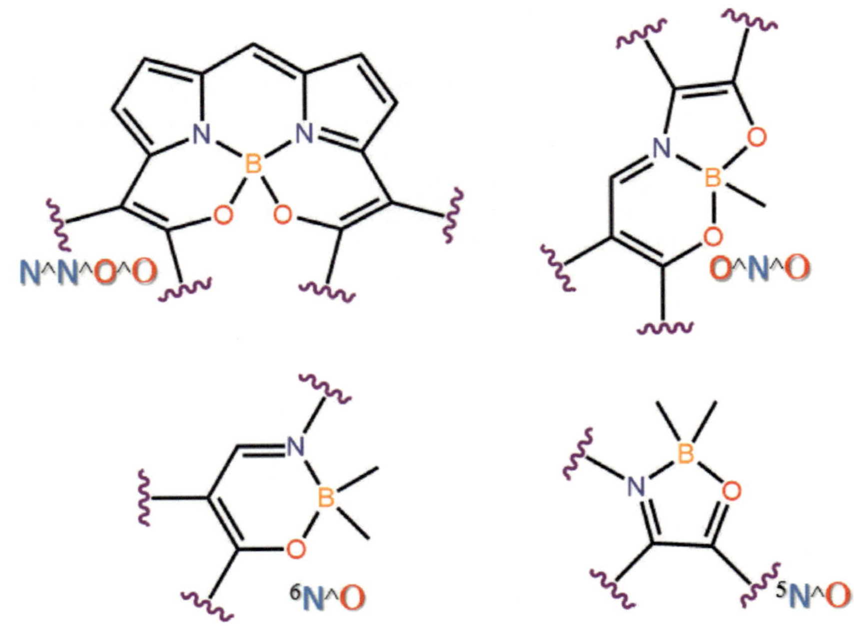

Figure 10.
Common type of chelating modes available to B(III) modules, as identified by Ziessel et al. [16].

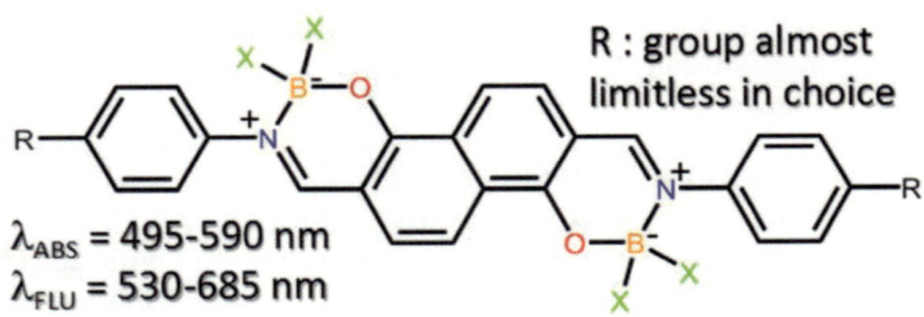

Figure 11.
Simple illustration of the construction of bis-boranil complexes around the rigid naphthyl central core. The nature of the terminal group can be varied at will while different functional groups can be attached to the boron(III) sites. The absorption and emission spectral properties can be modulated over a wide range.

the fluorescence quantum yields tend to be low, at least in solution, and thereby restrict the application of the compounds as solar concentrators. This is not the case for the rigid naphthyl-bridged binuclear complexes shown in **Figure 11** where emission quantum yields can approach 90% [17]. One added advantage of these simple chelates is the ability to synthesise optically resolved fluorophores in good yield. These latter compounds are rare among the boron(III) chelates in as much as their absorption and emission maxima, molar absorption coefficients and quantum yields are independent of the nature of the solvent.

5. Functional arrays

Returning momentarily to **Figure 1**, it can be seen that the array has a built-in element of redundancy to cover for damage to any particular chromophore. Even so, the terminal acceptor plays an important role and must be protected against damage. It must also fit tightly to the device in order for irreversible electronic energy transfer to be quantitative. The question raised by Lindsey et al. [9] concerns the ability of the array to deliver sufficient numbers of photons per unit time for the system to operate effectively. The device could be an organic solar cell or a molecular catalyst engineered for fuel production. In this latter case, the gap between arrival times of successive photons is crucial since the formation of stable chemical products requires multiple electrons. In order to avoid the damaging effects of free radicals, it is essential that charge accumulation is completed before the intermediate radical can escape and cause damage. In the case of a solar cell, the performance requires excitation with the optimum flux of photons. A single light-harvesting array of the type depicted in **Figure 1** is incapable of supplying the necessary input. The solution is to design systems where many individual arrays combine to furnish the device with sufficient photons. Since it is unlikely that a large number of arrays can be packed around the device in a logical manner, it follows that systems have to be designed whereby fast electronic energy transfer between arrays occurs with very high efficiency. Only one such system has been reported to-date [18] but the rate of energy transfer was too slow for practical operation (**Figure 12**). New design strategies are needed if progress is to be made in this critical direction.

A second concern about the viability of artificial light-harvesting antennae is the inevitability of chemical damage during prolonged exposure to sunlight. It seems impossible to completely prevent loss of chromophores during operation and, at present, we have no successful strategies for in-situ repair of damaged components. Approaches need to be found, therefore, to by-pass damaged units without compromising the overall antenna. This might be achieved using dyes that can be selectively protonated so as to induce a large change in the absorption spectrum when exposed to acidic conditions. There will be a significant alteration of the corresponding spectral overlap integral for electronic energy transfer and this simple modification will allow the direction of exciton migration to be controlled. This behaviour is illustrated by way of **Figure 13**, where the molecular antenna presented as **Figure 1** is re-examined.

Figure 12.
Pictorial representation of artificial light-harvesting array built by attaching disparate BOBIPY derivatives to a functionalized C60 residue. At high concentration in a thin plastic film, electronic energy migration proceeds between adjacent particles decorated with the yellow dye and exciton trapping occurs at the blue dye. This is a rare example of long-range energy transfer between particles. Reprinted with permission from [18]. Copyright (2012) American Chemical Society.

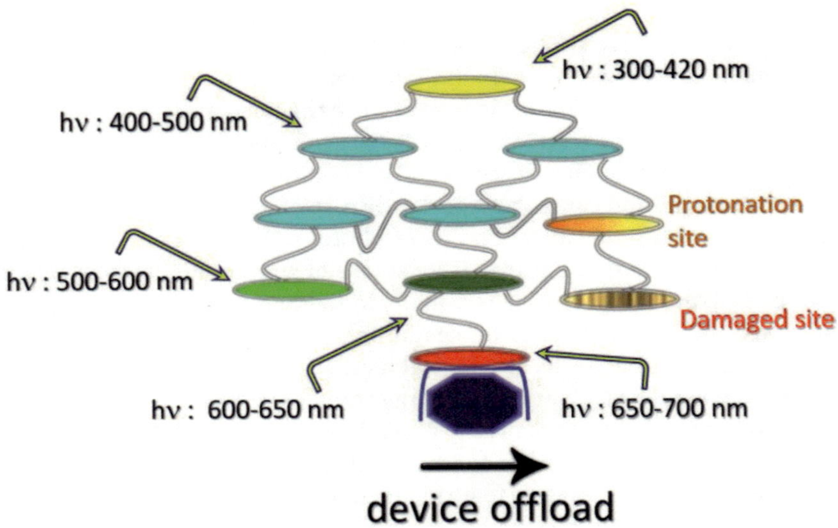

Figure 13.
*Modification of the generic set-up indicated in **Figure 1** to include the likelihood that one of the components will become damaged during prolonged exposure to sunlight. Such damage leads to release of a proton, which becomes attached to an adjacent chromophore and changes the absorption spectral profile. This in turn switches-off electronic energy transfer to the damaged site and re-directs the exciton within the antenna.*

It is presumed that under continuous illumination one of the components will become damaged. There are growing indications that such photofading of the chromophore follows autocatalytic kinetics, meaning that the system appears stable for a considerable period but suddenly starts to bleach. The product could catalyse further damage by, for example, formation of long-lived triplet states. Bleaching is often accompanied by release of a proton and this could be collected by an adjacent chromophore. Provided protonation leads to a marked change in colour, electronic energy transfer along that segment of the antenna could be interrupted and excitons directed elsewhere within the system. This leads to self-protection without loss of performance and avoids excitation of damaged components that might otherwise activate radical formation. **Figure 14** shows an example of a molecular triad that fulfils the desired objective [19]. Protonation can also be effected using a photo-acid activated with near-UV light.

One of the most important lessons learned from examination of the natural light-harvesting array is that, in all likelihood, successful artificial analogues will need to possess the facility to transfer excitons over hundreds of identical

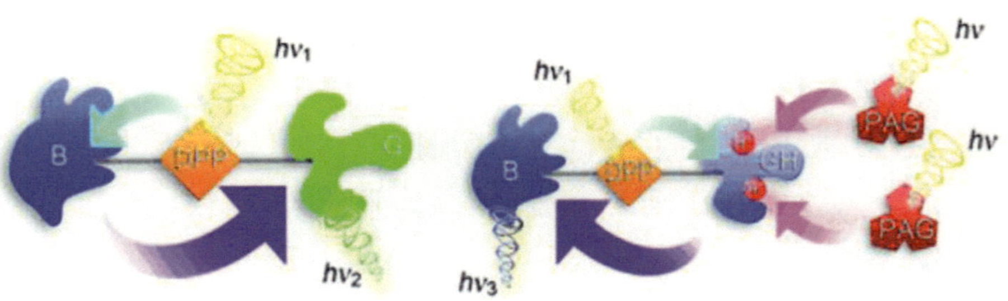

Figure 14.
Illustration of the use of a photo-acid to alter the direction of electronic energy transfer in a molecular triad. For the neutral molecule, photons absorbed by the central dipyrropyrole (DPP) unit are transferred rapidly to the green BODIPY terminal. This latter unit is protonated on illumination of a photo-acid generator (PAG). This changes the optical properties of the dye and, as a consequence, EET now occurs to the blue BODIPY dye. Reproduced with permission from [19].

chromophores. The direction of energy transfer cannot be controlled under such conditions because there is no driving force. Energy transfer will resemble a random walk. To be efficient, individual chromophores will need to be closely spaced but not so close that π,π interactions provide a route for fluorescence quenching. These realisations have led us to consider the use of photonic crystals as the basis for our artificial light harvesters. This might appear a rather bizarre strategy because it is well known that fluorescence quenching is usually very effective for high fluorophore concentrations in fluid solution due to self-association. There is the additional problem of self-absorption, which also tends to curtail fluorescence. None-the-less, crystals offer several key advantages relative to plastic films, most notably the very high absorbance that can be achieved. There are now several reports describing fluorescence from boron(III) chelates under visible light illumination.

Early work in this field noted that single crystals assembled from a BODIPY derivative equipped with a *meso*-tetrathiophene residue were fluorescent [20]. Selective excitation of the oligothiophene unit at 400 nm was followed by fast electronic energy transfer to populate the fluorescent state of the BODIPY dye. Such observations are important because they indicate close packing of the molecules in the crystal lattice does not necessarily restrict the photophysical properties of the molecular dyad. In other work, it was reported that BODIPY derivatives fitted with aryl groups in place of the conventional fluorine atoms formed crystals that were significantly more fluorescent than the same compound dissolved in fluid solution [21]. The crystal hinders rotation of the B-aryl groups and this closes down an important radiationless pathway. This is a further example of what is currently termed "aggregation-induced emission". The same crystals assemble into "columns" with adjacent units being co-facial, having a centre-to-centre distance of ca. 9.0 Å, but slightly offset and with the boron atoms lying on the same side of the molecule. Such arrangements could facilitate electronic energy migration along the column in a random walk (**Figure 15**).

Photonic crystals based on BODIPY derivatives contain an amorphous region in addition to the ordered columns. These somewhat disordered regions appear to facilitate dimerization, or higher-order aggregation, of the chromophore which is evident as a red-shifted absorption band. These self-associated species can operate as traps for excitons migrating along the ordered columns. This long-wavelength

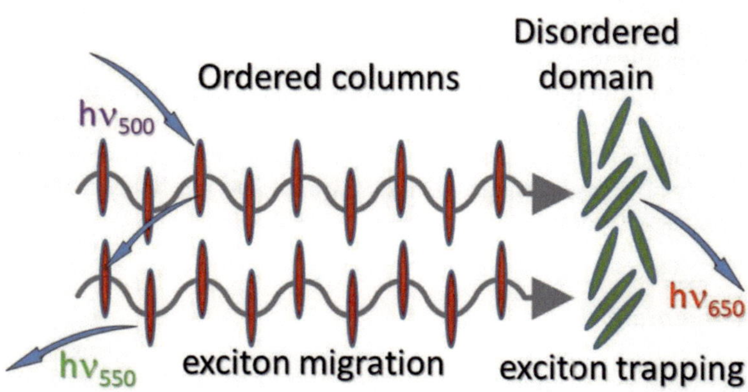

Figure 15.
Schematic representation of the photophysics that follow illumination of a BODIPY chromophore within the crystal lattice. The arrangement of molecules into layers provides an ideal environment for rapid electronic energy transfer along the layer. This process competes effectively with fluorescence such that an exciton might sample several hundred individual dye molecules. The exciton can hop between columns on a slower timescale. Aggregates form in a more amorphous region and, because of a red-shifted absorption spectrum, these act as traps for excitons migrating along the layers. It is quite rare for aggregates to fluoresce. Reproduced with permission from [21].

emission is in addition to the usual Frenkel-type fluorescence so that the overall fluorescence profile is broadened considerably. The same is true for the absorption spectrum. Our studies indicate that the aggregates can possess quite long-lived excited states and are therefore possible candidates as energy donors for opto-electronic devices. Constructing such systems represents the next phase of this project.

This work might pave the way for new types of artificial photon concentrators possessing high densities of chromophore. It is necessary to identify crystals where π,π-interactions are at a minimum and this could mean introducing steric blocking groups to keep individual molecules at some preferred distance. New experimental techniques will be needed to monitor the photophysical properties of single crystals, probably making use of fluorescent microscopes. However, by working closely with crystallographers, it should be possible to compile a set of rules regarding the best criteria for isolating photonic crystals suited for fast electronic energy transfer followed by strong fluorescence from a defect state. Such materials have many advantages compared to intricate covalent architectures.

6. Conclusion

In several respects, boron(III) chelates can be considered ideal candidates for building photonic arrays where many chromophores act cooperatively to harvest sunlight and to subsequently drive a useful device. The tetrahedral geometry around the boron atom favours the design of multi-component structures where the subunits are not in strong electronic communication. The same structural feature helps to isolate individual chromophores in the crystal lattice, despite the close proximity of neighbouring molecules. The basic macrocycle assembled around the boron(III) centre can be modified, expanded, diversified and functionalized with almost no limit, so as to provide access to an enormous range of compounds. Most of these materials are emissive, although it is necessary to avoid systems that are based on intramolecular charge-transfer processes. Fluorescence is often retained in the solid state and this is an unusual feature that can be well exploited to develop artificial photon concentrators.

This latter field has produced some wondrous examples of accreted molecular entities capable of highly efficacious electronic energy transfer within the cluster. The next stage of development requires groups of such clusters to operate together so as to amplify the number of photons concentrated at the device in unit time. This is a major challenge that demands the introduction of new types of synthetic methodologies. In principle, photonic crystals can solve this particular problem and can be doped at the surface with fluorescent traps. However, there remains the need to attach the crystal to the device without loss of performance of either unit. The field is only now beginning to fulfil its promise and it is likely that operational photonic crystals will soon emerge.

Acknowledgements

We thank Newcastle University for financial support of this work. HHTA-S gratefully acknowledges the award of a postgraduate scholarship from The Umm Al-Qura University (Saudi Arabia).

Conflict of interest

The authors declare no conflict of interest.

Author details

Hatun H.T. Al-Sharif and Anthony Harriman*
Molecular Photonics Laboratory, Newcastle University, Newcastle upon Tyne, United Kingdom

*Address all correspondence to: anthony.harriman@ncl.ac.uk

IntechOpen

© 2020 The Author(s). Licensee IntechOpen. This chapter is distributed under the terms of the Creative Commons Attribution License (http://creativecommons.org/licenses/by/3.0), which permits unrestricted use, distribution, and reproduction in any medium, provided the original work is properly cited.

References

[1] Fokas AS, Cole DJ, Chin AW. Constrained geometric dynamics of the Fenna-Matthews-Olson complex: The role of correlated motion in reducing uncertainty in excitation energy transfer. Photosynthesis Research. 2014;**122**:275-292. DOI: 10.1007/s11120-014-0027-3

[2] Marin A, Passarini F, van Stokkum IH, van Grondelle R, Croce R. Minor complexes at work: Light-harvesting by carotenoids in the photosystem II antenna complexes CP24 and CP26. Biophysical Journal. 2011;**100**:2829-2838. DOI: 10.1016/j.bpj.2011.04.029

[3] Loudet A, Burgess K. BODIPY dyes and their derivatives: Syntheses and spectroscopic properties. Chemical Reviews. 2007;**107**:4891-4932. DOI: 10.1021/cr078381n

[4] Ulrich G, Ziessel R, Harriman A. The chemistry of fluorescent bodipy dyes: Versatility unsurpassed. Angewandte Chemie, International Edition. 2008;**47**:1184-1201. DOI: 10.1002/anie.200702070

[5] Alamiry MAH, Benniston AC, Copley G, Elliott KJ, Harriman A, Stewart B, et al. A molecular rotor based on an unhindered boron dipyrromethene (Bodipy) dye. Chemistry of Materials. 2008;**20**:4024-4032

[6] Wagner RW, Lindsey JS. A molecular photonic wire. Journal of the American Chemical Society. 1994;**116**:9759-9760. DOI: 10.1021/ja00100a055

[7] Harriman A. Artificial light-harvesting arrays for solar energy conversion. Chemical Communications. 2015;**51**:11745-11756. DOI: 10.1039/c5cc03577e

[8] Harriman A, Stachelek P, Sutter A, Ziessel R. Stepwise photoconversion of an artificial light-harvesting array built from extended BODIPY units. Photochemical & Photobiological Sciences. 2015;**14**:1100-1109. DOI: 10.1039/c5pp00021a

[9] Hasselman GM, Watson DF, Stromberg JR, Bocian DF, Holten D, Lindsey JS, et al. Theoretical solar-to-electrical energy conversion efficiencies of perylene-porphyrin light-harvesting. The Journal of Physical Chemistry. B. 2006;**110**:25430-25440. DOI: 10.1021/jp064547x

[10] Tamgho IS, Hasheminasab A, Engle JT, Nemykin VN, Ziegler CJ. A new highly fluorescent and symmetric pyrrole-BF2 chromophore: BOPHY. Journal of the American Chemical Society. 2014;**136**:5623-5626. DOI: 10.1021/ja502477a

[11] Woodford OJ, Ziessel R, Harriman A, Wills C, Alsimaree AA, Knight JG. Optical spectroscopic properties recorded for simple BOPHY dyes in condensed media: The mirror-symmetry factor. Spectrochimica Acta. Part A, Molecular and Biomolecular Spectroscopy. 2019;**208**:57-64. DOI: 10.1016/j.saa.2018.09.047

[12] Woodford OJ, Ziessel R, Harriman A. Photofading of an extended BOPHY chromophore dispersed in poly(methyl methacrylate) as a chemical actinometer. ChemPhotoChem. 2018;**2**:1046-1054. DOI: 10.1002/cptc.201800130

[13] Huaulme Q, Mirloup A, Retailleau P, Ziessel R. Synthesis of highly fluorescent BOPHY chromophores displaying large Stokes shifts. Organic Letters. 2015;**17**:2246-2249. DOI: 10.1021/acs.orglett.5b00858

[14] Woodford OJ, Stachelek P, Ziessel R, Algoazy N, Knight JG, Harriman A. End-to-end communication in a linear

supermolecule with a BOPHY centre and N,N-dimethylanilino-based terminals. New Journal of Chemistry. 2018;**42**:4835-4842. DOI: 10.1039/c7nj04654e

[15] Maity P, Gayathri T, Singh SP, Ghosh HN. Impact of FRET between molecular aggregates and quantum dots. Chemistry, An Asian Journal. 2019;**14**: 597-605. DOI: 10.1002/asia.201801688

[16] Frath D, Azizi S, Ulrich G, Retailleau P, Ziessel R. Facile synthesis of highly fluorescent boranil complexes. Organic Letters. 2011;**13**:3414-3417. DOI: 10.1021/ol2011665

[17] Urban M, Durka K, Jankowski P, Serwatowski J, Luliński S. Highly fluorescent red-light emitting bis(boranils) based on naphthalene backbone. The Journal of Organic Chemistry. 2017;**82**:8234-8241. DOI: 10.1021/acs.joc.7b01001

[18] Iehl J, Nierengarten JF, Harriman A, Bura T, Ziessel R. Artificial light-harvesting arrays: Electronic energy migration and trapping on a sphere and between spheres. Journal of the American Chemical Society. 2012;**134**:988-998. DOI: 10.1021/ja206894z

[19] Hablot D, Harriman A, Ziessel R. Using a photoacid generator to switch the direction of electronic energy transfer in a molecular triad. Angewandte Chemie, International Edition. 2011;**50**:7833-7836. DOI: 10.1002/anie.201102065

[20] Benniston AC, Copley G, Harriman A, Rewinska DB, Harrington RW, Clegg W. A donor-acceptor molecular dyad showing multiple electronic energy-transfer processes in crystalline and amorphous states. Journal of the American Chemical Society. 2008;**130**:7174-7176. DOI: 10.1021/ja800387d

[21] Bozdemir OA, Al-Sharif HHT, McFarlane W, Waddell PG, Benniston AC, Harriman A. Solid-state emission from mono- and bichromophoric boron dipyrromethene (BODIPY) derivatives and comparison with fluid solution. Chemistry - A European Journal. 2019;**25**:15634-15645. DOI: 10.1002/chem.201903902

Chapter 4

Significance of Boron Nitride in Composites and Its Applications

Sasikumar Rathinasabapathy, M.S. Santhosh and Manivannan Asokan

Abstract

Boron nitride (BN) exists in several polymorphic forms such as a-BN, h-BN, t-BN, r-BN, m-BN, o-BN, w-BN, and c-BN phases. Among them, c-BN and h-BN are the most common ceramic powders used in composites to ensure enhanced material properties. Cubic boron nitride (c-BN) has exceptional properties such as hardness, strength than relating with other ceramics so that are most commonly used as abrasives and in cutting tool applications. c-BN possesses the second highest thermal conductivity after diamond and relatively low dielectric constant. Hence pioneer preliminary research in AMCs proven substitute composites than virgin AA 6061 traditionally used for fins in heat sinks. Moreover, poly-crystalline c-BN (PCBN) tools are most suitable for various machining tasks due to their unmatchable mechanical properties. h-BN also finds its own unique applications where polymer composites for high temperature applications and sp^3 bonding in extreme temperature and compression conditions.

Keywords: cubic boron nitride, metal matrix composites, mechanical properties, thermal properties

1. Introduction

The boron nitride conglomerate was initiated by Balmain 1842 and achieved to form powders by GE scientist named Robert H. Wentorf in 1957 only. GE named the product commercially as Borazon. During the launch period, it was costlier than gold in the market [1]. As such carbon, boron nitride also can possibly be produced in amorphous and crystalline forms. In translucent form, boron nitride occurs or forms in three allotropes: hexagonal boron nitride (h-BN) similar to graphite (**Figure 1A**), sphalerite boron nitride (β-BN) similar to cubic diamond, and wurtzite boron nitride (ɣ-BN) similar to hexagonal diamond form [2].

Exceptional to the carbon fullerenes (C_{60} buckyballs), BN fullerenes have most common bonding nature of squares or octagons than pentagons to avoid adverse thermo-dynamical properties due to unfavorable B–B and N–N bonding (**Figure 1D**) [3]. Similar to one-dimensional carbon nano tubes (CNTs), BN nano tubes (BNNTs) (**Figure 1C**) are also isoelectric to CNTs with the correlation to tube diameters, chirality, and numerous numbers of walls [4]. Among different BN forms, h-BN is the steady and stable phase of BN and interest initiated succeeding the graphene sheets isolation in 2004 The structural texture of h-BN is layered structure and inside layer of each, the boron atoms and atoms of nitrogen are bound

IntechOpen

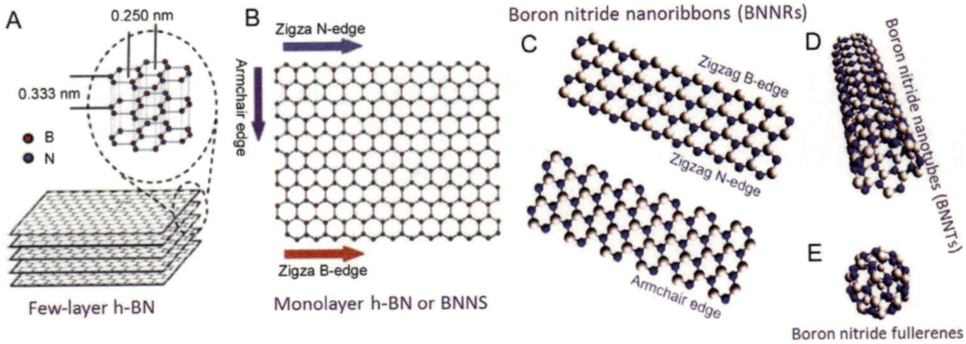

Figure 1.
Boron nitride allotropes: (A) few-layer hBN, (B) h-BN monolayer nanosheet or BNNS, (C) boron nitride nanoribbons (BNNRs) with various edge cessation leading to zigzag B-edge and structures of armchair edge, (D) boron nitride nanotubes (BNNT), and (E) BN fullerenes. Reproduced with the permission from "The Royal Society of Chemistry."

strongly due to covalent bonds nature in-plane and forces of van der Waals held together at each layer. A h-BN single layer is typically termed as BN nanosheet or BNNS. This formal structure is possible only for h-BN sheets since small aspect ratio [5]. For materials with higher aspect ratio their widths typical measurement will be <50 nm, they are known as BN nanoribbons or BNNR's. Even h-BN has similar graphene structure, yet material's bandgap is wide with 5.9 eV intrinsic band gap comparing more conductive graphene. Since its conductive thermally, h-BN is suitably fascinating for various electronic applications. h-BN possibly be used as filler material which insulates electricity for thermal radiators, polymer or ceramic composites, UV emitters and field emitters. Exceeding its properties of insulation, h-BN is inert chemically among wide variety of solvents, acids, and oxidizers. It is also insoluble in usual acids but soluble in nitrides (Li_3BN_2) and molten alkaline salts (LiOH, KOH). By high resistance nature chemically and thermal stability behavior, it finds as a fascinating material in hazardous environment [6].

2. Structural properties of h-BN

It is a powder (white in color and slippery nature) in physical form, in other way analogous to graphite. The size of the flake commercially available are h-BN varying from hundreds of nanometers to tens of microns. Thus, BNNS sheets developed through exfoliation process with these crystals are very frequently restricted to the maximum lateral sizes (few tens of microns) of the initial initiating material. Individual h-BN monolayers or BNNSs are structure with fluctuating atoms of boron and nitrogen combined to form a honeycomb. The bond length of B–N is 1.45 Å and B–N is covalent bond type. The neighboring borazine centers rings distance is 2.5 Å. The BNNS edge structure can be zigzag or armchair, alike graphene. The antecedent is a B or N-edged structure, in contrast hindmost is a BN pair-edged structure. The h-BN crystal structure is hexagonal with $P6_3/mmc$ space group (**Figure 1A**), lattice constants, a = b = 0.2504 nm, c = 6661 nm, bond angles, $\alpha = \beta = 90°$, $\gamma = 120°$. The BN is partially ionic structure in h-BN which reduces conductance capacity of electricity, covalence and dissimilar to graphite, it benefits the AÀ stacking sequence [7]. In most cases, this is the best energetically beneficiable stacking sequence noticed in BN, amongst them B atoms (electron-deficient) are directly above or below the N atoms (electron-rich) in the adjoining layer. The easiest possibility to calculate the number of h-BN layers by observing from transmission electron microscopy (TEM) images of folded edges. **Figure 2A** and **B** depicts TEM images of high-resolution, which usually gives atomic lattice information more along with the layers in number. An additional simple proficiency to analyze/calculate h-BN thickness by atomic force microscopy (AFM).

The number of layers is possibly determined by observing the step height usually with respect to the surface (**Figure 2D** and **E**). Mostly sheets of monolayer h-BN formed by means of chemical vapor deposition process (CVD) and mechanical exfoliation process have a approximate height of 0.4 nm. In chemical exfoliation sheets, the height measured can vary as maximum to1 nm due to the solvent trapped in the middle of the substrate and h-BN flakes. Thus, HRTEM generally used to report number of layers, while recording the height by means of AFM and more characterization techniques needed to ensure the findings. Another convenient characterization technique and simple method, by means of an optical microscope aimed to locate the amount of layers in graphene sheets and it is also used for h-BN moreover. Normally, ~300 nm of a SiO_2 layer coated standard silicon wafer (**Figure 2C**) is benefited as a substrate for reference and the layers in number is determined depends on the noticed optical contrast using the respective microscope. In h-BN case, as absorption not takes place in visible region, the oxide layer's

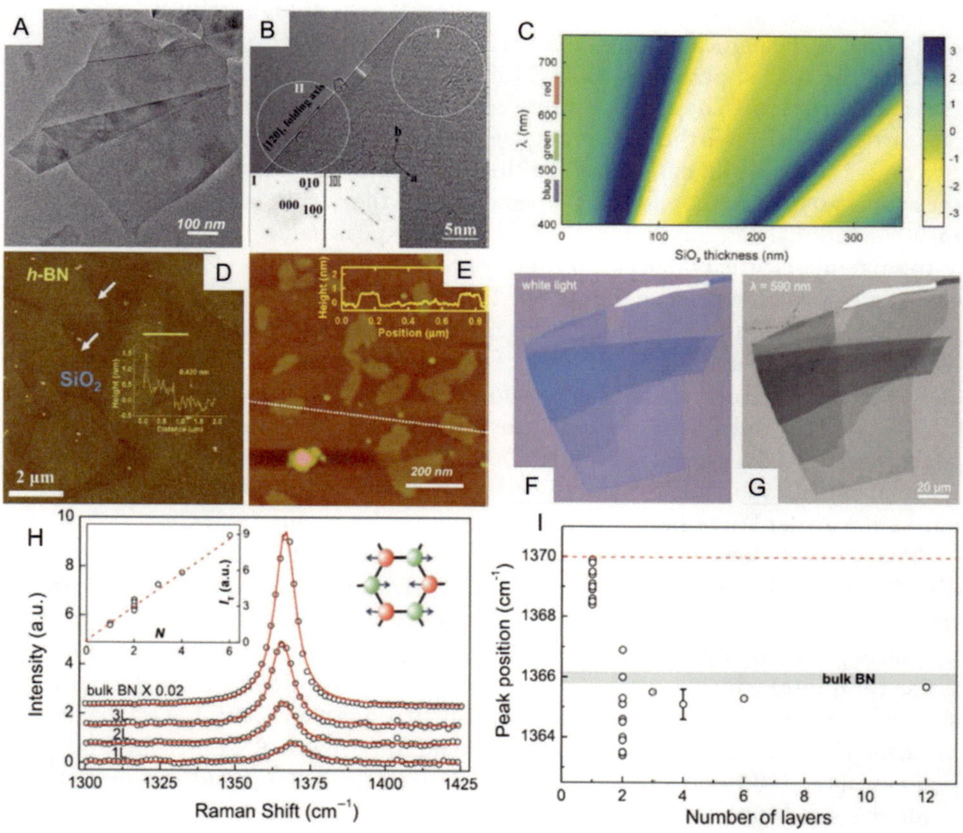

Figure 2.
(A) Low-exaggeration TEM image of an exfoliated BNNS. (B) HR-TEM image displaying the layers number at the overlap edges (Lin et al., 2010a). (C and D) Height map from AFM of CVD fatten BNNS displaying the single layer 0.4 nm thickness (E) chemically exfoliated BNNS displaying a 1 nm height (F) white light microscopy image of exfoliated h-BN flake displaying the optical contrast on 90 nm SiO_2/Si. (G) Enhanced h-BN optical image flake using a light source of 590 nm displaying the various contrast in-between layers. (H) BNNS in Raman spectra to a few-layer h-BN in relating the variance in the peak width to number of layers. (I) Top position difference as observed for single layer to BNNS bulk. Reproduced with the permission from corresponding author—Gorbachev (2011).

substrate need to be altered for 80 nm in spick and span to optically recognizing the single to few-layered materials. As flaunt in **Figure 2F**, a low value of 2.5% reaches the contrast white light in h-BN monolayer, that is, tetra times lesser than graphene, and the increases in contrast with the various number of layers increase in the same [8]. In other way, for identifying the amount of layers different wavelength light is being used, which can be seen in **Figure 2C**. By using ~590 nm wavelength light as flaunt in **Figure 2G**, the noticed contrast between numbers of layers was clearly obvious. This layer recognition study be further proceeded to Raman microscopy. Vibrational mode of boron nitride noticed at 1364–1371 cm^{-1}, which depends purely on the number of layers. **Figure 2H** and **I** flaunt the relative study by Gorbachev et al. relation to BNNS different layers. It is possible to be viewed in **Figure 2H** for monolayer, drifts to higher wave number of 1370 cm^{-1} noticed due to the Raman peak broadens and when it compared with the bulk, which is 1366 eV. This study evidently flaunts BNNS formation through the red shift, which could be used later as standard to ensure BNNS presence [9].

3. Thermal stability

Allotropes of BN, in-particular h-BN and c-BN exhibit high stability in terms of thermal and chemical nature. h-BN is more stable even without deteriorating at exceeding the temperatures 1000°C (air), 1400°C (vacuum), and up to 2850°C (inert atmosphere). The h-BN theoretically noticed thermal conductivity values are close to grapheme. It is the one among best materials which are thermally conductive that is available till date. Based on the structure of BNNR, at room temperature the arm chair edged ribbons are 20% smaller than the zigzag-edged BNNRs. In-plane thermal conductivity determined as high peak value of 390 W/m K even at room temperature, which is 280 times greater than that of the silicon dioxide, for electronic devices which generates heat deciding h-BN as an attractive material due to its dielectric nature. Alike graphite, because of its anisotropic strength bonding, h-BN also strongly exhibits coefficient of thermal expansion (CTE) anisotropically. The coefficient of thermal expansion (CTE) in the a-direction (in-plane) is $\sim 2.90 \times 10^{-6}\ K^{-1}$ during room temperature, in the c-direction, when the CTE is 10 times larger and over to value of $4.05 \times 10^{-5}\ K^{-1}$ at room temperature. The greater thermal expansion in positive note along the c-direction is mainly because weak nature in van der Waals bonding in the middle of planes [10].

The popularly accepted boron nitride phase diagram explains various boron nitride phases was determined from thermodynamic properties. At ambient conditions, it is known that c-BN is thermodynamically stable rather than h-BN. But this is contrary to carbon phase diagram at ambient conditions; example is hexagonal phase (graphite) which is a more stable phase. The boron nitride phase diagram is shown in **Figure 3A** and **B**, where original calculations were indicated by the dashed lines and the refined diagram indicated by solid lines. The h-BN/c-BN/liquid triple point exists at 3480 ± 10 K and 5.9 ± 0.1 GPa, while the h-BN/liquid/vapor triple point exists at 3400 ± 20 K and 400 ± 20 Pa.

Based on the phase diagram above temperatures below 1600 K, it is obvious that c-BN is more favorable than h-BN. The general temperatures for growth in CVD are between 700 and 1100°C, which is 300°C at least cooler than that noticed in the above phase diagram. This odd behavior can be due to Gibbs free energy prevailing with the system. The temperature of transition travels in the middle range of h-BN and c-BN shifts as a Gibbs free energy function (Shift ± 10 meV/atom). The transition temperature varies between 1200 and 1800 K due to the change in free energy; mainly for the h-BN growth the typically lowest value used. There may be small

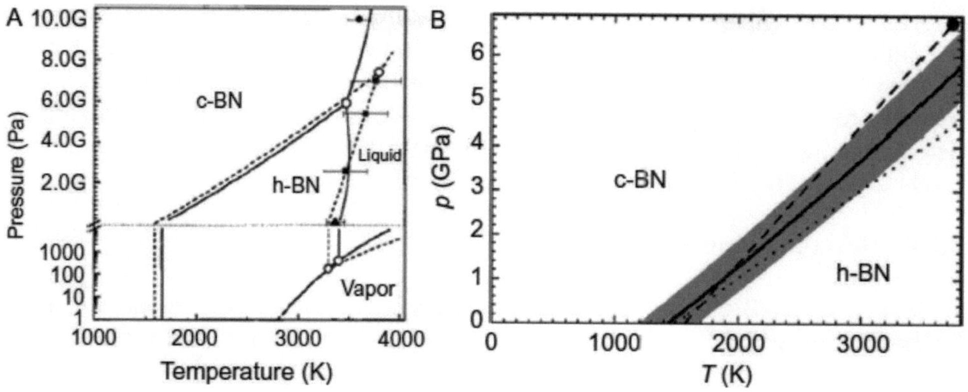

Figure 3.
(A) h-BN phase diagram displaying the stability nature of h-BN vs. c-BN at different temperatures and pressures. (B) Zoom image of the transition phase and Solozenko calculations (dotted line) (Solozhenko et al., 1999).

discrepancies casually caused by defects, grain size, contaminants, or sometimes interactions with the substrate transition metal itself [11].

4. Applications for h-BN

2D h-BN are considered greatly for use in various applications including substrate dielectrics for high-temperature resistive layers anticorrosive coatings, nanoelectronic devices, frictional layered coatings and also other 2D material systems and gate which are termed as advanced applications. The large bandgap (~6 eV), chemical resistivity and high temperature both the direct-grown h-BN and exfoliated h-BN materials are ideal for various hazardous applications like chemical and thermal industry.

4.1 Dielectrics for future generation nanoelectronic devices

Graphene is a well known for its highest electron and also whole mobility nature of any material till date, for future generation high-speed electronic gadgets and devices this 2D material a natural candidate. Initially graphene devices were transferred onto traditional substrates of SiO_2 and noticed a reduction in inherent transport properties because of its contrary reciprocal action with substrate (>120,000 cm^2/Vs compared to device performance on SiO_2 of ~2000 cm^2/Vs). The transporter potency in the devices was diminished because of the scattering from charged surface states and also due to impurities in SiO_2, the comparatively more roughness in surface of the substrate, and also the very low-frequency optical modes of phonon in surface of SiO_2. Dean et al. [12] was the team initiated study bulk h-BN's use as a substrate material to restrain these outcomes seen in SiO_2 devices. The h-BN's substrate strong in-plane bonding was suspected to permit for an atomically smooth, inert surface free from dangling bonds, or charge traps in the surface. This was trusted to be true, as the graphene's electronic properties increased including improved potency of transportation, decreased in transportation due to in-homogeneity, and decreased inherent intoxicating from the substrate were appreciably enhanced. The h-BN integrated first exfoliated GFET in the gate dielectric and also in the substrate, desired result obtained when manufacturing h-BN/graphene/h-BN sandwiched device. When related with a conventional GFET device produced along with SiO_2 and Al_2O_3, as dielectrics, a 70% enhance in the peak transconductance (gm) was noticed with a higher operational frequency as well as appreciable than 5X enhance in graphene's potency of transportation from 1200 to 6500 cm^2/Vs. h-BN is a vital part in graphene and other 2D material-based electronic devices as a gate dielectric. In dielectric at a constant value of 6, voltage breakdown value of 8–10 MV/cm, and nearing bandgap value of 6 eV which are advantageous over typical SiO_2 dielectrics can be accomplished [13].

4.2 Vertical tunneling device and behavior

In inclusion to BN as a substrate material for graphene used electronics, few layer BN has been examined as a high-quality, low-dielectric constant railing material for 2D rampant electron tunneling devices where different compositions electrodes are divided by the dielectric layer thickness. These devices depend on the excavating via the ultrathin crystalline films layers and allow the smallest allowable gate lengths down to a single atom, and also permitting for exceptionally fast transport.

Excavating measurements via few-layer, device using exfoliated h-BN which constructs have been effectuated on a substrate which is conductive using conductive type atomic force microscopy (C-AFM). With beneficial tip areas normally on the 103 nm^2 order, fundamental characteristic merits of dielectric able to be examined and that are non dependent on general failure mechanisms of whether macro- or either microscopic dielectric including dust particles, grain edges and cross-plane defects.

4.3 h-BN in protective coatings

While h-BN research much concentrated on nanoelectronic applications, technologies of others use the ultrathin and impactful studies on layered materials also done. Thin h-BN used for corrosion resistance and antioxidation protective coatings, in particular, it is a applied science that got validated by chemical inertness and h-BN's stability even at high temperature to efficiently emblem the underlying material from revelation to a numerous elements. Due to inert nature and lack of swinging loosely bonds, h-BN possibly be an excellent antifriction coating, alike many other two dimensional materials which share same characteristics.

4.4 h-BN in gas sensing

h-BN also been used as a essential substance to sense gases such as ammonia and ethanol. h-BN atomic layers were used for manufacturing gas sensors and concentration of about very low (100 ppm) amount of gas was allowed to pass across the device. The technology in this mechanism is as follows: When sensor is air exposed, the absorbed oxygen molecule gets ionized through available free electrons of h-BN conduction band, the output is an enhance of the resistance. Identically, when ethanol gets introduced, it effects a resultant action on the surface, i.e., decreasing the resistance because of electrons releases back to the sensor. Thus, this mechanism proves the efficiency even at varied temperatures and concentrations of ethanol.

5. Synthesis of c-BN

In 2015, Caldwell et al. explored the utmost case of a-BN as initiating material. They concluded a-BN possibly be permuted into c-BN at above 7.0 GPa pressures and temperature of above 1070 K. On condition that portal pressure is bigger than the required in starting material which is of poorly crystalline h-BN, in contrast the portal temperature is controlled beneath by about 400 K. A transformation accentuates the chances of two mechanisms relative to kinetic study. First is direct crystallization transformation process from a-BN to c-BN and the next is two step-process which forms h-BN in an in-between stage: a-BN--" h-BN ~ c-BN. When t-BN is dealt as the initiating material, in which (B, N) layers along the c axis are randomly distributed, c-BN's very small crystallites only present in operating conditions of temperature and pressure close to 6 GPa and 1250 K. In collateral to the investigation of eminent researchers it was concluded that, operating conditions of temperature and pressure given phase change figure for BN, it is feasible to balance c-BN at higher temperatures under the c-BN-h-BN equilibrium line provided by Wentorf. The transformation of h-BN--" c-BN stands on certain important factors such as structural excellence of starting material purity of used products etc. In many of the instances its importance's are not still defined significantly.

The professed catalysed process is the common method used for deriving c-BN at industrial scenario. Certain amalgams added to h-BN in the process of decreasing the higher activation energy barricade are mostly termed as catalysts or solvents. These terms are not exact to the required; because this initiating product does not act as a motivational element but preferably act as a flux antecedent, giving eutectic h-BN. In this eutectic, BN is partially dissolved. In this process, h-BN's starting material is either closely mixed or only stay in touch with flux antecedent. The impetus for formation of c-BN is the solubility difference between h-BN and c-BN varieties in the eutectic flux, beneath the fixed conditions of high pressure and temperature. In the (P, T) territory of c-BN balance, c-BN is impulsively nucleated and can grow faster. In spick and span to form p- or n-type semiconductors, doping of c-BN crystals possibly be done by impurities introduced into the lattice of zinc blende using high purity starting materials. After caring and curing process at high pressure and temperature, the c-BN particles recuperated are separated by particular chemical or physical methods (e.g., acids and/or molten alkalis—usually molten $NaOH$-Na_2CO_3 mixtures—are used) in spick and span to remove the flux antecedents and by-products as well as non-reacted h-BN residues [14].

6. Cubic boron nitride: synthesis methods

6.1 At static high pressure condition: crystal growth

The temperature variance method process includes the BN transport from hot area containing h-BN to a cold zone (where one or more c-BN seeds are placed) by using a solvent. The time needed for growth of crystal is higher than time handled for the synthesis as described early and it possibly reach several days. The flux antecedents used are generally those of group A materials. Yazu et al. patented a pressure cell in which the pattern of work consists of varied chambers splitted up by non-reactive layers. In conjunction to get good quality crystals, researchers selected flux antecedents as Li_3BN_2, $Mg_3B_2N_4$, $Ca_3B_2N_4$, $Sr_3B_2N_4$, $Ba_3B_2N_4$ or combinational constituents of these compounds. For example, good quality crystals can form after 60 hours, a 0.25 carat (50 mg) crystal at 5.5 GPa in-between a range of 1890–1540 K temperature gradient. The temperature variance can be changed by means of two main methods in the middle of nutrient and seed (i) by varying the position of a growth cell in a vertical furnace, minding the presence of a temperature gradient in-between the center and the top of the heater tube or (ii) by thrusting a thin molybdenum sheet with a drill hole, also as a baffle, in-between the flux and the h-BN source and changing the diameter of this hole. These two techniques allow control of the super saturation which is responsible for the difference noticed in the crystal growth shape. The crystal shapes can change from polyhedral to dendritic geometries as the super saturation increases [15].

6.2 Dynamic high pressure process

This technique is mainly used for compacting c-BN powders. The emerging movement of a boron nitride impenetrable form produced under shock compression was studied by Sawaoka and Akashi. By single shock compression, they noticed only the transformation to w-BN with a greater decrement in the size of particle (by a factor of 10) in relation with the initiating material. A compression of type double-shock leading to form an amorphous material and a very little quantity (amount) of c-BN (certain per cent). These particles of c-BN are 2D seeds. The synthesis of c-BN requires shocks in multiple during compression. If w-BN is capitulated to a shock compression at above 10 GPa pressure, c-BN can be derived. Sawaoka and Akashi proposed the following sequence to achieve c-BN transformation. By single-shock treatment, a greater amount of lattice strains in numerical exists in the w-BN crystals and these flaws are enhanced with subsequent second shock during compression process. This is mainly because of the energy variance in-between wurtzite-type and zinc blende-type stacking sequence is very little quantity, yet the strains in lattice are enough to accomplish the w-BN-* c-BN conversion. Sato et al. process patented for producing c-BN from r-BN by shock wave compression. Dremin et al. studied the crack in the frail dominion region and the velocity of particle for trinitrotoluene-RDX blends with supplements of h-BN. They reviewed the works in which signs of the "weak detonation regime" have been found [16].

7. Cubic boron nitride: industrial and potential applications

7.1 Mechanical applications

c-BN, in relation to hardness correlating to diamond, is much more harder than the traditional abrasive materials like Al_2O_3, SiC and boron carbide. Hence, the output performance of grinding with c-BN wheels are enhance increased over the traditional abrasive material (SiC or Al_2O_3) wheels during grinding hardened high speed steels and cast irons of chilled mill grade. Tools made out of diamond cannot generally be used in above cases due to its chemical reactivity nature with ferrous metals, whereas c-BN is against to chemical attack in ferrous metals existence up to 1500–1600 K.

For cutting tools, c-BN is used as sintered c-BN. The c-BN sintering should be done in the region of thermodynamic stability with the pursuit to prevent retransformation into h-BN at higher temperatures while at the process of sintering. Alike diamond, c-BN is hard-to-sinter material typically, due to its bonding nature of strong covalence and its stableness at higher operating conditions of temperature and pressure. There are possibly two processes: the instantaneous sintering of c-BN while the conversion of h-BN or two-step sintering. During the previous case, the preliminary step is the transformation from h-BN to c-BN (utilizing a traditional higher temperature and pressure flux transformation method) exemplified by the isolation and purification of c-BN in powder form and the next step involves sintering during rest or movable during high pressure conditions.

Hirano et al. concluded work on the one-step process as follows; the concurrent sintering and transformation of c-BN from h-BN by means of adding A1N. The operating conditions of pressure and temperature are same as that of c-BN formation. It is observed that Fukunaga used additives to enhance transformation and sintering which is 2 mol.% lesser than that of magnesium nitride (Mg_3N_2). He observed the complete change or an conversion during 5.0 GPa and 1770 K for 1 hour. Fukunaga and co-workers used Mg_3BN_3 as sintering agent and formed semi-transparent c-BN polycrystals, at the working conditions of 6 GPa pressure and 1770 K as temperature. The formed c-BN polycrystals are with 99% of the calculated density.

Akashi and Sawaoka studied shock compaction about the sintering of c-BN powder. The powders without any form of additives are directly handled using shock compression. The starting powders must be coarse. The density and micro hardness of the produced c-BN compacts are dependent in nature strongly based on the size of the starting powder. They obtained compacts with 98% of theoretical density and micro hardness of 51.3 GPa. Shintani et al. compared the microstructural metallographic study and the properties of mechanical associated with two sintered c-BN materials. Singh reviewed sintered materials of super hard nature such as diamond, w-BN and c-BN. By examining them through X-ray diffraction, scanning electron microscopy (SEM) and energy-dispersive X-ray analysis.

In most cases, the compacted c-BN powder is cemented on tungsten carbide or ceramic substrates. Sintered cutting tools allow ferrous metals, chilled cast irons and hardened steels to be machined. The cutting speed is augmented and life of the tool is increased influentially by the value of 7–50 (in comparison with tungsten carbide tools). The high quality of the surface finish dispenses with further grinding and polishing actions in many cases [17].

7.2 Applications in electronics

c-BN is an III-V semiconductor compound. Its very wide band gap makes it a very good insulator. Furthermore, c-BN has highest thermal conductivity, which allows it to be used as a heat sink for semiconductor lasers, microwave devices etc. In these applications, the surface of c-BN ceramics is coated with a group VIII metal or aluminium (or aluminium alloy) by CVD process. It is possible to metalize as innovated by Tanji and Kawasaki with nickel by CVD process or with gold or aluminium by sputtering.

c-BN is able to be mixed with silicon and also with beryllium in pursuance to get p-type or n-type semiconductors respectively. Mishima et al. patented a growing method semi conductable c-BN crystals. For example, when $LiCaBN_2$-Si mixture used as a flux antecedent to the h-BN-" and c-BN transformation at 5.5 GPa and at a temperature of 2070 K for 18 hours time period, researchers acquired n-type c-BN crystals size by 1.2 mm. These doped crystals of c-BN able to produce p-n junction diodes that even works at high temperatures due to c-BN's higher thermal stability. Injection scintillation in the ultra violet was noticed at a high pressure from a c-BN made p-n junction. This emission of light occurs near the certain region especially at junction only based on several conditions. Certain methods of producing p-n junctions from c-BN semiconductors formed through a high pressure, high temperature processes have been patented. Recently, Ahmad and Lichtman have studied c-BN thin films for UV sensor applications [18].

The c-BN electronic applications have only begun to be developed; their use should be increased significantly in the future. Because of the small size of the components required in electronics, compounds or thin films with a higher value of thermal conductivity and electrically insulating properties are needed [12].

8. Conclusion

The boron nitride amalgamation combination was noticed by Balmain in the year 1842 and synthesized to powders by Robert H. Wentorf a GE scientist at 1857. Boron nitride occurs in various polymorphic forms like a-BN, h-BN, t-BN, r-BN, m-BN, o-BN, w-BN and c-BN phases. h-BN exists in three types of allotropes. BNNTs are isoelectric in nature to CNTs. h-BN is attractive for various electronic applications due to its thermal conductive behavior. It also can be used as filler materials in thermal radiators, UV emitters and field emitters. h-BN and c-BN shows high stableness both thermally and chemically. Direct grown h-BN and exfoliated h-BN materials are ideal for hazardous applications such as thermal and chemical industry. Exfoliated h-BN is favorably feasible with conductive type atomic force microscopy (C-AFM). h-BN can also be used as an excellent antifriction coating, corrosion resistance and also as antioxidation protective coatings. h-BN also used in gas sensing at any varied temperatures and concentration. There are two significant methods for cubic boron nitride synthesis; at static high pressure crystal growth observed and dynamic high pressure process. Sintered c-BN is necessary for mechanical applications to prevent conversion to h-BN at higher temperatures. c-BN powder is cemented on tungsten carbide tools and which allows ferrous metals, chilled cast irons and hardened steels to be machined. The doped crystals of c-BN are able to produce p-n junction diodes which can even works at higher temperatures due to its higher thermal stability. c-BN electronic applications have recently started to get developed; their significant usage will be developed in mere future. The future researchers can pay attention on the above area as well as on metal matrix composites in particular aluminium to produce better efficient and durable materials for modern era and requirements.

Acknowledgements

This chapter is extracted from the sponsored research project which is sponsored by Science and Engineering Research Board (SERB), Government of India. As a Principal Investigator of this research project, I wish to thank the SERB and also the Management Selvam College of Technology for their consistent motivation & support to bring out this chapter successfully.

Author details

Sasikumar Rathinasabapathy[1*], M.S. Santhosh[2] and Manivannan Asokan[2]

1 Selvam Composite Materials Research Lab, Department of Mechanical Engineering, Selvam College of Technology, Namakkal, Tamilnadu, India

2 Research Scholar, Selvam Composite Materials Research Lab, Department of Mechanical Engineering, Selvam College of Technology, Namakkal, Tamilnadu, India

*Address all correspondence to: sasikalipatty@gmail.com

IntechOpen

© 2019 The Author(s). Licensee IntechOpen. This chapter is distributed under the terms of the Creative Commons Attribution License (http://creativecommons.org/licenses/by/3.0), which permits unrestricted use, distribution, and reproduction in any medium, provided the original work is properly cited.

References

[1] Alem N, Erni R, Kisiel Owski C, Rossell M, Gannett W, Zettl A. Atomically thin hexagonal boron nitride probed by ultrahigh-resolution transmission electron microscopy. Physical Review B. 2009;**80**:155425

[2] N. No. Manufacture of Boron Compounds, United States Patent Office; 1958. pp. 2-4

[3] Anota EC, Tlapale Y, Villanueva MS, Ma'rquez JAR. Non-covalent functionalization of hexagonal boron nitride nano sheets with guanine. Journal of Molecular Modeling. 2015;**21**:215

[4] Balmain WH. Bemer kungen uber die Bildung von. Verbindungen des Bors und Siliciums mit Stickstoff und gewissen Metallen. Journal für Praktische Chemie. 1842;**62**(1):422-430

[5] Barth JV, Costantini G, Kern K. Engineering atomic and molecular nanostructures at surfaces. Nature. 2005;**437**:671-679

[6] Bhaviripudi S, Jia X, Dresselhaus MS, Kong J. Role of kinetic factors in chemical vapor deposition synthesis of uniform large area graphene using copper catalyst. Nano Letters. 2010;**10**:4128-4133

[7] Constantinescu G, Kuc A, Heine T. Stacking in bulk and bilayer hexagonal boron nitride. Physical Review Letters. 2013;**111**:036104

[8] Bresnehan M, Hollander MJ, Wetherington MT, Wang K, Miyagi T, Pastor G, et al. Prospects of direct growth boron nitride films as substrates for graphene electronics. Journal of Materials Research. 2013;**29**:1-13

[9] Bresnehan MS, Bhimanapati GR, Wang K, Snyder DW, Robinson JA. Impact of copper overpressure on the synthesis of hexagonal boron nitride atomic layers. ACS Applied Materials & Interfaces. 2014;**6**:16755-16762

[10] Caldwell JD, Kretinin A, Chen Y, Giannini V, Fogler MM, Francescato Y, et al. Sub-diffraction, volume-confined polaritons in the natural hyperbolic material, hexagonal boron nitride. Nature Communications. 2014;**5**:1-9

[11] Caldwell JD, Vurgaftman I, Tischler JG. Probing hyperbolic polaritons. Nature Photonics. 2015;**9**:638-640

[12] Dean CR, Young AF, Meric I, Lee C, Wang L, Sorgenfrei S, et al. Boron nitride substrates for high quality graphene electronics. Nature Nanotechnology. 2010;**5**:722-726

[13] Coleman JN, Lotya M, O'Neill A, Bergin SD, King PJ, Khan U, et al. Two dimensional nano sheets produced by liquid exfoliation of layered materials. Science. 2011;**331**:568-571

[14] Corso M, Auw€arter W, Muntwiler M, Tamai A, Greber T, Osterwalder J. Boron nitride nanomesh. Science. 2004;**303**:217-220

[15] Cortes CL, Newman W, Molesky S, Jacob Z. Corrigendum: Quantum nano photonics using hyperbolic meta materials (2012 J. Opt. 14063001). Journal of Optics. 2014;**16**:129501

[16] Golberg D, Bando Y, Huang Y, Terao T, Mitome M, Tang C, et al. Boron nitride nanotubes and nanosheets. ACS Nano. 2010;**4**:2979-2993

[17] Cui Z, Oyer AJ, Glover AJ, Schniepp HC, Adamson DH. Large scale thermal exfoliation and functionalization of boron nitride. Small. 2014;**10**:2352-2355

[18] Dai S, Ma Q, Andersen T, Mcleod AS, Fei Z, Liu MK, et al. Sub diffractional focusing and guiding of polaritonic rays in a natural hyperbolic material. Nature Communications. 2015a;**6**:6963

Chapter 5

Hexachlorinated Boron(III) Subphthalocyanine as Acceptor for Organic Photovoltaics: A Brief Overview

Georgy L. Pakhomov, Vlad V. Travkin and Pavel A. Stuzhin

Abstract

A boron(III) complex of peripherally hexachlorinated subphthalocyanine, Cl_6SubPc is a very promising small-molecule acceptor for application in organic photovoltaics. In this chapter the recent experimental results in the field are compared, and a critical review is given of the published works on the solar cells with the planar or bulk heterojunction architectures. The thin film properties of Cl_6SubPc are also considered. The approaches to the further modification of the molecular structure of boron(III) subphthalocyanine-type compounds for the enhancement of their photoelectrical properties are discussed.

Keywords: boron(III) subphthalocyanines, acceptors, organic electronics

1. Introduction

Recent achievements in the synthetic chemistry of subphthalocyanine-type compounds have led to an appearance of numerous molecular structures in a variety of shapes with markedly different redox and optical properties [1–3] and durability. However, only a few of those structures deserve the close attention of engineers involved in organic electronics.

The thin film electronic devices utilizing both subphthalocyanines and diverse phthalocyanine-type compounds are traditionally classified as organic light-emitting diodes (OLEDs), organic field-effect transistors (OFETs), and organic photovoltaic cells (OPVC) [1, 4–7]. Of course, these fascinating heterocyclic molecules find their way into other areas of application such as resistive memory or molecular switches [8, 9], but those are incomparably inferior to the above three types of thin film devices in terms of practical achievements.

The light-emitting properties of conventional phthalocyanines in the undoped films do not appear to be very promising, so they are included in the OLED scheme mostly as the charge transporting components [1, 4, 10, 11]. Plint et al., who are continuing a long-term research in this field, proposed two SubPc derivatives as dopant emitters for an (8-hydroxyquinolinato)$_3$Al matrix to obtain the white light from a standard multilayer OLED made by the high-vacuum sublimation technique [12]. Also worth mentioning here is the earlier work of Torres' group [13], in which the solution-processable OLED structures incorporating variously substituted SubPcs were fabricated and tested.

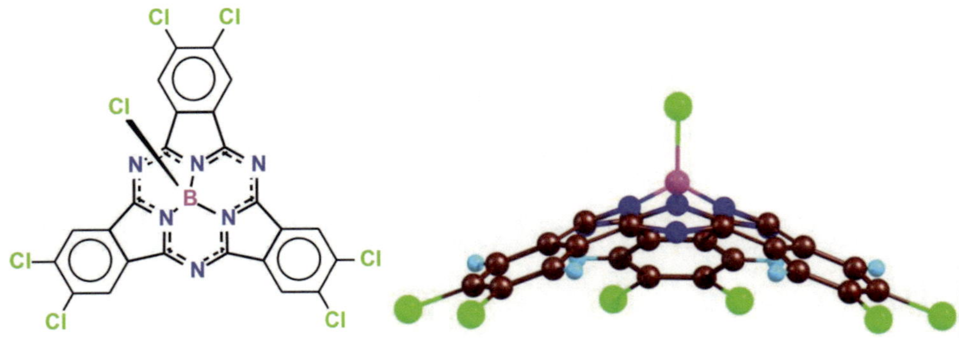

Figure 1.
Cl_6SubPc molecule: the planar chemical structure with π-conjugation (left) and the computed 3D structure illustrating the conical geometry (right). Green balls show chlorine atoms, one in the axial position (extraligand) and six on the periphery of the benzene rings.

As to the applications in OFETs, the readers should turn to the review in Ref. [14]. In short, the charge carrier mobilities measured using the standard transistor geometry in subphthalocyanine films are lower than in the films of four-leaf phthalocyanines that are capable of forming highly ordered structures in the channel. A noteworthy work was published more than 10 years ago by Yasuda and Tsutsui [15] who observed sign inversion (from N- to P-) in the majority charge carriers when transferring the SubPc-based OFETs with gold S/D electrodes from the glove box to ambient air. This adds intrigue to the current discussion of whether SubPc acts as a P- or N-type layer in the junction with another molecular material: a possible reason for that may be the unintentional doping from the adjacent layers or the unevenness of fabrication/measurement conditions for different prototypes.

The most impressive results today appear in the field of photovoltaics. We therefore will discuss SubPcs only from the viewpoint of their behavior in thin film photovoltaic cells, particularly in cells with an organic or hybrid heterojunction based on the hexachlorinated subphthalocyanine derivative Cl_6SubPc (**Figure 1**). This compound is gaining practical importance, though many other peripherally halogen-substituted SubPc derivatives have been synthesized over the last decade [1–5, 16]. To the best of our knowledge, only one hexachlorinated derivative with an extraligand other than chlorine was used in a photovoltaic cell [17]. Despite the availability of many sound publications that deal with Cl_6SubPc as the photovoltaic material, little is known about its fundamental (intrinsic) properties in a solid. In addition, we have found some inconsistencies in the discussions and speculations unsupported by the relevant experimental data.

2. Properties of Cl_6SubPc thin films

After the introduction of electron-withdrawing chlorine atoms on the periphery of a macrocycle, the Cl_6SubPc molecule behaves as the electron acceptor with respect to the molecule on the other side of the heterointerface (e.g., unsubstituted SubPc [6, 18–22]), thereby participating in the separation of the photogenerated charges. In the bulk phase, addition of chlorine atoms to the conjugated macrocycle assists in getting the electron transporting organic semiconductors [7]. Therefore, judging from its chemical formulae (**Figure 1**), Cl_6SubPc is a priori assumed to be the accept or N-type material for the organic photovoltaic cells. Perhaps, it is due to this assumption that too little effort has been made to fundamentally characterize N-type conductivity in the Cl_6SubPc solid, except for a series of electro-physical experiments on the vacuum-deposited thin films, carried out by Beaumont et al. in [23].

The energy level alignment at SubPc/Cl_6SubPc heterojunction was recently discussed in detail [21] with the emphasis on the interface gap $E_G = E^{Donor}{}_{HOMO} - E^{Acceptor}{}_{LUMO}$ that determines the maximum achievable open-circuit voltage in a photovoltaic device with a D/A junction. The interface gap E_G of 1.95 eV was derived from soft XPS and XAS, with the use of DFT (**Table 1**) [21]. This value of E_G is much greater than what is attainable through pairing the SubPc donor with the conventional N-type acceptor C_{60}, which explains the popularity of Cl_6SubPc among the so-called "non-fullerene" acceptors for organic photovoltaics [5, 16, 17, 21, 25, 27, 30, 32, 33]. The DFT calculations on a Cl_6SubPc molecule were also performed in Refs. [6, 33], including the visualization of the most important frontier MOs. However, even in the abovementioned studies, such parameters as transport

Parameter	Value	Ref.	Comment
Sublimation temperature, °C	270–290	[24]	Greater than for many $halogen_xSubPc$
Color	Purple	[24]	In powder and in toluene solution
$\lambda_{max}(Q)$, nm,	576	[25]	In chloroform
In solution	569	[24]	In toluene
	570	[26]	In dichloromethane
In sublimed film	585–590	[18, 19, 24]	Identical with SubPc
E_{opt}, eV	2.1	[20]	Link to Ref. [28]
	2.16	[27]	Not specified
	2.1	[28]	From $\lambda_{max}(Q)$ in solid film
	2.19	[24]	From $\lambda_{max}(Q)$ in solution
E_{act}, eV	1.38	[24]	From thermally activated conductivity in thin vacuum-deposited films
HOMO, eV[a]	6.69	[6]	From UPS
	5.9	[20, 23]	Adopted from [28]
	6.0	[21]	From XPS, XAS, and DFT
	5.8	[22]	From cyclic voltammetry
	6.06	[27]	From cyclic voltammetry
	6.0	[28, 29]	Adopted from [22]
LUMO, eV[a]	3.8	[20, 23]	Adopted from [28]
	3.9	[21]	From XPS, XAS, and DFT
	3.7	[22]	From HOMO-E_{opt} difference
	3.61	[30]	From cyclic voltammetry
	3.90	[27]	From HOMO-E_{opt} difference
	3.6	[28, 29]	From HOMO-E_{opt} difference
Charge carrier mobility μ, $cm^2 V^{-1} s^{-1}$ at RT	$8 \pm 2 \times 10^{-7}$	[23]	Vertical zero-field electron mobility, from impedance spectroscopy
	8.3×10^{-6}	[27]	Blend with PTB7-Th, vertical electron mobility, from SCLC
	7.6×10^{-5}	[31]	Charge carrier mobility of the sum of holes and electrons Σ_μ, from flash-photolysis time-resolved microwave conductivity

[a]*These are actually negative, below vacuum level.*
Methods: UPS = ultraviolet photoelectron spectroscopy, XPS = X-ray photoelectron spectroscopy, XAS = X-ray absorption spectroscopy, DFT = density functional theory, SCLC = space-charge-limited current.

Table 1.
Cl_6SubPc, some experimental data.

gap, exciton binding energy E_{bi}, or diffusion lengths L_D, which are of paramount importance when constructing a heterojunction-based device [6, 34], have not been estimated[1].

Table 1 summarizes the data on some solid-state properties of Cl_6SubPc available in the literature. As expected, the most frequently discussed values are the positions of the highest occupied and lowest vacant molecular orbitals (HOMO and LUMO) since they can be used for deriving E_G and, further, for constructing the band diagrams that predict the basic photoconversion parameters of the devices employing various donor/acceptor pairs [20–23, 28–30, 35].

As seen from **Table 1**, the HOMO-LUMO values obtained by various methods/research groups fluctuate within the range of ±0.2 eV. Moreover, different numbers can be reported in the articles by the same authors ([20] *vs.* [28] or [21] *vs.* [22]). Such discrepancies are not surprising, considering the diversity of approaches/methods used for estimation [36]. Certain caution should therefore be taken when comparing the data from different sources and, especially, when relying on them in the interpretation of the device characteristics [30, 36]. For instance, in the same Ref. [30], E_G for the SubPc/Cl_6SubPc pair was calculated to be equal to 1.83 eV, i.e., the value did not coincide with that from the Ref. [21]. Importantly, the optical bandgap E_{opt} is smaller than the fundamental bandgap (the difference between the ionization potential and the electron affinity of a molecule) by the value of the electron–hole pair binding energy. Thus, the HOMO-LUMO difference obtained from the quantum chemical calculations with various levels of accuracy provides an approximation to the fundamental gap and needs to be amended before used as equivalent of E_{opt} (**Table 1**). The situation gets more complicated in a solid, where the transport of charge carriers or quasiparticles (excitons) occurs. Due to the polarization/stabilization effects, E_{bi} differs from the binding energy of electron–hole pair created within a (single) photoexcited molecule in the gas phase [36].

Cl_6SubPc sublimes in a high vacuum without noticeable decomposition, which allows one to obtain thin films on various functional substrates. The vacuum-deposited films are intensively purple colored, homogeneous, and smooth [24]. This favors their application in multilayer photovoltaic devices. However, they are electrically resistive [21, 24, 35], with the charge carrier mobility (electrons) much lower than in fullerenes (**Table 1**). In the literature, the Cl_6SubPc films are commonly treated as amorphous [21]. As recently shown, a correct choice of deposition surface and temperature contributes to the production of crystalline films of Cl_6SubPc with improved conductivity, which in turn upgrades the parameters of photovoltaic devices [24, 35]. To our knowledge, the effect of intentional or unintentional (e.g., atmospheric) doping of the Cl_6SubPc matrix on its conducting properties has not been investigated so far.

3. Fabrication of Cl_6SubPc-based heterojunctions for photovoltaic cells

Roughly, there are two types of heterojunctions with clear geometrical dissimilarity: flat heterojunction (PHJ) and bulk heterojunction (BHJ). Cl_6SubPc can be used in both—Refs. [5, 18–23, 25, 28–30, 35] and [5, 6, 27, 33], respectively. Such flexibility is accounted for by the increased, as compared to phthalocyanines, solubility of Cl_6SubPc, and subphthalocyanines in general, although a rigorous study of

[1] We have found only one source [Barito AJ. Cascade Organic Photovoltaics [thesis]. University of Michigan; 2015] unsupported by the relevant peer-reviewed journal publication, in which L_D = 4.5 nm and exciton lifetime τ = 0.53 ns are reported for the vacuum-deposited Cl_6SubPc films.

the solvation processes does not permit unambiguous attribution of conventional SubPc to the classical dyes, it is pigment-like rather [26, 37].

PHJ is prepared by sequentially depositing thin layers of donor and Cl_6SubPc. The most common A-on-D order of deposition gives a normal PHJ, assuming the cathode located on top of the device, while the D-on-A configuration with the anode on top is termed as inverted PHJ. The thickness of layers is of paramount importance, and it should be adjusted individually not only for the donor but also for Cl_6SubPc, since the experimental setup, deposition sequence, additional functional materials, etc. may vary in each research group. Thicker photoactive layers are likely to capture more incident photons and protect against leakages, but short L_D and rapidly increasing serial resistance require that the layers be kept sufficiently thin. Typical thickness of Cl_6SubPc in PHJ does not exceed few tens nanometers. Deposition process is carried out mostly via the vacuum evaporation technique, but the examples of solution-processed PHJ with subphthalocyanines are also known [38, 39].

To obtain a BHJ, a donor component and Cl_6SubPc should be dissolved in the chlorobenzene (various additives like 1,8-diiodooctane (DIO) or 1-chloronaphthalene (CN) are recommended) and then deposited using the spin-coating technique. Both the solution chemistry and post-deposition treatment of thus made blend affect the resulting device parameters [17, 27, 33].

It is generally believed that the efficiency of power conversion (*PCE*) in the photovoltaic cells with BHJ must be greater than in the PHJ-based cells, this being due to the morphological reasons [38–40]. Quite surprisingly, in all the works dealing with both solution- and vacuum-processed BHJ with a Cl_6SubPc acceptor, the actual morphology of the photoactive layer(s) remains highly uncertain. In Ref. [6], the morphology of 20 nm thick films consisting of a co-evaporated in vacuum $SubPc_x$:Cl_6SubPc_{1-x} blend has not been elucidated. In Ref. [27], the polymer: Cl_6SubPc mixture termed as BHJ was spin-coated and annealed to obtain 75 nm thick photoactive layers (**Table 2**), but the transmission electron microscopy (TEM) images showed a homogeneous blend without a noteworthy phase separation. This led authors to a conclusion that the pure phase domains are absent. Notably, the nanocrystalline domains with high local carrier mobility of at least one of the two components of BHJ are required for efficient dissociation of the charge transfer (CT) states into free charge carriers at the D/A interface [27].

A broader morphological study in Ref. [33] included atomic force microscopy (AFM), TEM, and X-ray diffraction (XRD) measurements on the spin-coated mixtures of substituted subphthalocyanines, including Cl_6SubPc, with a polymer, also termed as BHJ. However, a thorough examination of the published images reveals that neither AFM nor TEM indicates a formation of a long-enough fibrillary structure in the films (unspecified thickness), although authors stated otherwise[2]. The XRD patterns display the changes in the intensity of a single diffraction peak corresponding to the polymeric component of the blend relative to the pure phase of the polymer (at only one D/A ratio) [33], which cannot be regarded as a proof of BHJ.

In neither of the above studies, a relevant morphological model has been developed based on the instrumental analysis. The "phase diagram" attributing the morphological changes to the composition of a binary phase [43–45] is missing as well. The well-known "mosaic" picture of the polymer: subphthalocyanine BHJ published in 2009 [38] seems to be borrowed from the sketches of the polymeric solution-made BHJ [16, 45] without being confirmed by adequate morphological analysis. Contrary to what is drawn, the authors stated at the end of the article [38] that the films stayed amorphous even for the 1:5 blend (from grazing incidence

[2] Authors claimed in the supplementary materials section that their Cl_6SubPc-based cells outperform the BHJ-based cells fabricated in Ref. [27], which is not true (**Table 2**).

Ref.	Schematic of cells*	Junction type	Parameters			
			J_{sc} (mA/cm^2)	V_{oc} (V)	*FF*	*PCE* (%)
[18, 19]	A1	PHJ	3.10	1.33	0.59	2.39
	A2		3.62	1.29	0.47	2.20
	A3		3.28	1.32	0.63	2.70
[22]	B	PHJ	3.53	1.31	0.58	2.68
[23, 28]	C1	PHJ	2.54	0.89	0.50	1.12
	C2		2.09	0.50	0.48	0.50
[30]	D1	PHJ	6.17	1.00	0.66	3.96
	D2		10.1	1.04	0.67	6.86
[25]	E1	PHJ	9.0	1.03	0.71	6.4
	E2		9.0	1.02	0.68	6.1
	E3		9.2	1.03	0.58	5.4
[28, 29]	F1	PHJ	5.72	0.60	0.56	1.89
	F2		2.13	0.44	0.57	0.52
[35]	G	PHJ	6.43	0.55	0.48	1.71
[27]	H	BHJ	10.7	0.77	0.48	4.0
[33]	K	BHJ	7.79	0.66	0.48	2.48
[41]	L	PHJ/ BHJ	8.6	0.92	56	4.46
Cells with SubNc as acceptor or donor						
[42]	M	PHJ	14.55	0.96	61	8.40
[39]	N1	BHJ	10.3	0.90	41	3.8
	N2		12.1	0.74	47	4.2

Description of the schematic (thickness of layers in nm is given in the parentheses, if specified by authors):
A1 = ITO/MoO$_x$(5 nm)/SubPc(15)/Cl$_6$SubPc(20)/BCP(8)/Al, normal; A2 = ITO/BCP(8)/Cl$_6$SubPc(15)/SubPc(15)/MoO$_x$(40)/Al, inverted; A3 = ITO/MoO$_x$(5)/SubPc(14)/Cl$_6$SubPc(30)/BCP(5)/Al, optimized; B = ITO/MoO$_x$(5)/SubPc(10)/ Cl$_6$SubPc(27)/BCP(8)/Al; C1 = ITO/MoO$_x$(5)/Tc(60)/Cl$_6$SubPc(35)/BCP(8)/Al(100); C2 = ITO/MoO$_x$(5)/Pent(60)/Cl$_6$SubPc(25)/BCP(8)/Al(100); D1 = ITO/MoO$_3$(5)/SubNc(14)/Cl$_6$SubPc(8)/BCP:C$_{60}$(50)/Ag; D2 = ITO/PEDOT:PSS/DIP(5)/SubNc(14)/Cl$_6$SubPc(8)/BCP:C$_{60}$(50)/Ag; E1 = ITO/PEDOT:PSS/DIP(5)/SubNc(12)/Cl$_6$SubPc(10)/BCP:C$_{60}$(45)/Ag; E2 = ITO/PEDOT:PSS/DIP(5)/SubNc(12)/Cl$_6$SubPc(10)/C$_{60}$(35)/BCP(10)/Ag; E3 = ITO/PEDOT:PSS/DIP(5)/SubNc(19)/Cl$_6$SubPc(10)/BCP:Yb(45)/Ag; F1 = ITO/PEDOT:PSS/α6T(55, evaporated)/Cl$_6$SubPc (20)/BCP(10)/Ag(100); F2 = ITO/PEDOT:PSS/P3HT(55, spin-coated)/Cl$_6$SubPc(20)/BCP(10)/Ag(80); G = ITO/MoO$_x$(3)/CuI(60)/Cl$_6$SubPc(45)/BCP(6)/Al(100), structured; H = ITO/ZnO(40)/PTB7-Th:Cl$_6$SubPcCl (~75, solution)/MoO$_x$(10)/Ag(100); K = ITO/PEDOT:PSS(40)/PBDB-T:Cl$_6$SubPcCl(?, solution)/Ca(20)/Al(100); L = ITO/MoO$_x$(10)/DPSQ(13)/C$_{60}$(15)/C$_{60}$:ZCl:Cl$_6$SubPc(60,2:1:1 co-evaporated blend)/BCP(10)/Al(100); M = ITO/PEDOT:PSS(20)/α6T(60)/SubNc(12)/SubPc(18)/BCP(7)/Ag(120); N1 = ITO/MoO$_3$(5)/SubNc:PC$_{70}$BM(75,1:5 solution)/BCP(6)/Al(100); N2 = ITO/MoO$_3$(5)/SubNc:C$_{70}$(75,1:5 co-evaporated blend)/BCP(6)/Al(100).
Donors: SubPc = unsubstituted subphthalocyanine; Tc = tetracene; Pent = pentacene; SubNc = unsubstituted subnaphthalocyanine; α6T = α-sexithiophene; P3HT = poly(3-hexylthiophene); PTB7-Th = polymer with linear formula $(C_{49}H_{57}FO_2S_6)_n$; PBDB-T = polymer with linear formula $(C_{68}H_{78}O_2S_8)_n$; DPSQ = 2,4-bis[4–(N,N-diphenylamino)–2,6-dihydroxyphenyl] squaraine.
Other device components: ITO = indium-tin oxide, In_2O_3:SnO_2; MoO$_x$ or MoO$_3$ = molybdenum trioxide, usually substoichiometric; PEDOT:PSS = poly(2,3-dihydrothieno-1,4-dioxin)-poly(styrenesulfonate); BCP = bathocuproine; DIP = diindenoperylene ($C_{32}H_{16}$), CuI = cuprous iodide, ZnO = zinc oxide; ZCl = chlorinated zinc dipyrrin; PC$_{70}$BM = [6,6]-phenyl-C$_{71}$-butyric acid methyl ester.

Table 2.
Parameters of the photovoltaic cells with Cl$_6$SubPc as acceptor.

wide-angle X-ray scattering) and that "work is under way to improve the crystallinity of SubPc derivatives" [38].

Therefore, the researchers rely only on the fact that a film containing a mixture of two components is a BHJ. But rigorously speaking, there are certain morphological features that determine the ultrafast charge transport—the reason behind the success of the BHJ concept in organic photovoltaics [45]. The most important of them is the formation of bicontinuous interpenetrating 3D networks of components within a D/A blend. These networks, often thought of as columnar (pipelines), comb-like, or interdigital structures, should accomplish the transport of photogenerated charge carriers to the respective electrodes. Obviously, breaking the continuity will lower the charge collection efficiency. There is ample experimental evidence of the existence of such interdigitation in the classical solution-processed polymer: fullerene blends obtained by a set of independent, complementary analytical techniques, including the secondary ion mass spectrometry (SIMS) with depth profiling [45, 46].

We have found out that the misconception about any two-component layer as being truly a BHJ rests on the analogy with few examples of the photovoltaic cells, in which the (sub)phthalocyanine is mixed with C_{60}. Vacuum-deposited SubPc:C_{60} blends are widely known in organic photovoltaics due to availability and high efficiency; they are often used as a model system in many theoretical and experimental works. However, there is only one work where their microstructure was sufficiently detailized. Pandey et al. [44] conducted a comprehensive characterization of thin films of a mixture of SubPc:C_{60}, involving XRD, TEM, optical spectroscopy, and selected area electron diffraction. At only one mixing ratio of 1:4, they found some signs of crystallinity of SubPc dispersed in the nanocrystalline C_{60} matrix, whereas other compositions resulted in amorphous films. The *PCE* of the photovoltaic cells with thus optimized composition of the photoactive layer was only 9% higher than in a similar cell, in which the composition was not optimized (1:9). Besides, the morphology characterization and photovoltaic measurements were carried out on different substrates and films of different thickness, which requires further refinement[3].

Even for the metal-phthalocyanine complexes that tend to self-assemble (crystallize) due to a strong intermolecular interaction, a reliable experimental observation of the anticipated nanostructuring in co-evaporated blends turned out to be very problematic. As-deposited phthalocyanine: fullerene mixtures (commonly used ratio is 1:1) adopt fully amorphous structure, as witnessed via a combination of several independent analytical techniques [40, 51–55]. Again, it points to a lack of the phase separation, which is a prerequisite toward the formation of the interpenetrating D/A network typical of BHJ [45]. Only through considerable effort, by thoroughly selecting the annealing temperature, mixing ratio, or seed layer, the microscale organization of a blend can be initiated, and the pure crystalline phthalocyanine domains become visible [44, 52–54]. Unlike phthalocyanines, the SubPc molecules weakly interact in a solid, which makes their self-assembly rather difficult. As such, the formation of a well-established charge carrier percolation pathways in the SubPc:C_{60} blend will be less probable than in the phthalocyanine-based blends.

Crystallographic studies show that interactions between the neighboring Cl_6SubPc molecules in a solid are stronger than those in SubPc [23], but the formation (and observation) of a well-organized BHJ incorporating subphthalocyanine-type acceptor is still a challenging task.

---

[3] In mid 1990s, studies of the photoconductivity in thick films of the C_{60}-doped zinc phthalocyanine revealed the formation of a charge transfer complex that amplifies the photosensitivity of the blend [47, 48]. This was confirmed in Refs. [49, 50] using several optical methods, but later the authors sided with another model more closely associated with the BHJ, again with no morphological indications. In either way, the formation of a bimolecular CT complex means that the uniformly 1:1 mixed phase cannot be treated in terms of individual organic semiconductors any longer. Here, it is worthwhile to look further into the matter by proposing the new insights on the photoconductivity mechanisms instead of pursuing adaptation of the standard polymeric BHJ concept to the amorphous small-molecule based blends [51].

In any heterojunction type, the charge transfer across the D/A interface could also be influenced by the dipole moment of a molecule of one or both components. It bound to occur in the pyramidal subphthalocyanine molecules bearing axially attached halogen (**Figure 1**) and can further be tuned by the axial/peripheral substitution [4, 26, 56–59]. Mutual orientation of the acceptor/donor molecules at the interface affects both the dissociation distance and the local electric field during the charge transfer and separation, thus modifying the resulting V_{oc} of the device [59]. Morris et al. [59] experimentally investigated and modeled the characteristics of PHJ-based photovoltaic cells with two subphthalocyanines containing either chlorine or fluorine extraligand paired with the C_{60} acceptor. These donor molecules have nearly identical structure, except for a permanent electrical dipole, which allows the analysis of the variations in V_{oc} in terms of D/A separation width, polaron pair binding energy, and dipole orientation, other morphological factors being neglected.

Unfortunately, this interesting issue has not been given enough attention in the experiments. Theoretical considerations regarding the influence of a molecular dipole in a series of, mostly hypothetical, subphthalocyanines on the photovoltaic parameters were recently published [58]. The first-principles DFT calculations were also performed in [60] to characterize the electronic structure of the axially substituted SubPc molecules interfaced with C_{60}. A strong correlation between the experimentally measured V_{oc} and the computed CT excited state energy was found. One concluding remark hints that the dependence of these parameters on the actual interface morphology can be greater in significance than the modification of the ionization potential induced by change in the chemistry. Another prediction is that to gain a higher value of V_{oc}, the D/A interaction should be lowered, e.g., by increasing the spatial separation through the introduction of steric hindrances [60]. A fundamental theoretical study of relative arrangements of the donor and acceptor molecules was carried out by a large group of authors using the pentacene/C_{60} system as an example [61]. It was shown that the interfacial dipole originates mostly in polarization effects rather than a partial charge transfer from donor to acceptor. Next, the calculations demonstrate that the measurement of the macroscopic dipole averaged over the interface is not a representative of the local dipoles that can be induced by individual molecules at the interface. The local dipole was found to fluctuate in sign and magnitude over the interface and appears as if a sensitive probe of the relative arrangements of the pentacene and C_{60} molecules.

Note that theoretical findings are usually done under the assumption of an inert, molecularly sharp, regular, and pure PHJ (which almost never occurs in the experiments) and lack systematic verification in a representative series of the prototypical devices. For instance, the experimental evidence on the formation of a Diels-Adler adduct at the pentacene/C_{60} interface was recently found [62], thus casting doubt on the above results.

These and many other morphological issues are also addressed in the monograph, Chapter 2 in Ref. [63]. This book in general is strongly recommended for the readers interested in organic photovoltaic devices and materials.

4. Analysis of the cell performance

Table 2 summarizes the efficiency metrics for the prototypical solar cells exploiting Cl_6SubPc adopted from various sources, along with a description of the device schematics. For comparison, a few examples of the cells with subphthalocyanine are included in the last two rows. Other data collections describing the performances of variously designed photovoltaic devices based on SubPc-type compounds can also be found in Refs. [1, 4, 5, 28, 64].

As seen from **Table 2**, the single-junction photovoltaic cells with Cl_6SubPc as the acceptor can generate open-circuit voltage V_{oc} in the range of 0.44–1.33 V, usually about 1 V or above. Expectedly, the decisive contribution in the resulting *PCE* makes the short-circuit current J_{sc}; its value varies several times among different devices, while fill-factor *FF* lies within a range of 0.57 ± 0.12. The reported parameters largely depend on the device scheme, adjacent functional materials, and measurement conditions used in a particular study, which make their comparison difficult. Nonetheless, there are several points to ponder on when looking at the collected data:

1. Pairing Cl_6SubPc with the congener donors, SubNc or SubPc, results in the photovoltages exceeding 1 V. This value is greater than ever reported for the analogous PHJ utilizing structurally dissimilar small-molecule acceptors, like perhalogenated phthalocyanines or perylenes. Presumably, the structural similarity of the pairing conical molecules allows them to form a more intimately bound D/A pairs at the interface. This would mean, for example, a good quality of the (less distant) physical contact at the P/N junction. Second, as discussed above, certain orientations of electrical dipole of the donor molecule relative to the intrinsic dipole of acceptor, if one exists, could favor the charge separation. Macroscopic interfacial dipole at the SubPc/Cl_6SubPc junction was found to be small (0.15 eV) in [21], but in principle its role can be significant [51, 57, 59, 61, 63][4].

2. The devices obtained entirely by the vacuum deposition techniques usually exhibit better characteristics than devices with the solution-processed heterojunction (**Table 2**). At least two reasons could be suggested, taking into account that in reality the metallic contacts and oxide buffer layers are vacuum-evaporated even in the cells referred to by the authors as "solution-processed." First, the combination of wet and dry laboratory techniques used for the growth of multilayered heterostructures incurs problems with the compatibility of materials, transfer of semi-finished samples to the evaporator and back, etc. Using the vacuum methods only, the fabrication of the entire sample can be realized within a single run without breaking the growth process, from etching of substrates to deposition of the top electrode (including characterization tools, most of which require high vacuum). Second, the simplicity and robustness of the solution-based deposition techniques are somewhat overrated in the case of BHJ based on small-molecules. This is illustrated in the preceding section that describes the difficulties in obtaining the phase-separated bicontinuous networks with SubPcs. The BHJ concept does not provide the expected benefit in efficiency and is particularly unuseful for improving V_{oc}.

3. The champion efficiency of 8.4% reported for the PHJ-based photovoltaic cells so far has been obtained for the α6T/SubNc/SubPc cascade [42], which geometrically is a sequence of vacuum-evaporated PHJs. Here, both SubNc and SubPc behave as acceptors with respect to the thiophene molecules while being donors when paired with fullerenes or halogenated subphthalocyanines (**Table 2**). The excitons freely migrate across the relatively thick layers from

[4] In Ref. [65], a complex study of the "copper phthalocyanine/C_{60}" interface by DFT, UPS, and SIMS suggests that the local net charge-induced electric field, rather than the spontaneous charge transfer across the interface, is responsible for the interface dipole, in accordance with the theoretical predictions [61]. Authors observed a sizable interface dipole of electrostatic nature (up to 0.27 eV, depending on molecular orientation with respect to the deposition surface and on the deposition sequence), which rules out the charge transfer as the origin of the interface dipole.

the wide-bandgap to the smaller-bandgap acceptor with a subsequent dissociation at the donor interface via a long-range Förster energy transfer, which is in turn a function of the relative orientation of the transition dipole moments and distance between the molecules. The mechanisms explaining why both acceptors actively contribute to the photocurrent caused an active response in the organic photovoltaic community and spawned a large number of emulations. The attempt of Bender et al. [20] to use Cl_6SubPc as the donor component of the cascade cell with the scheme "Cl_6SubPc/μ-oxo-SubPc/C_{70}" was unsuccessful, highly likely due to the wrongly cascading LUMOs (but a very small layer thickness permitted enough efficiency). The cascade cells in Refs. [25, 30], although the authors do not consider them as such, demonstrated a high photovoltaic performance (**Table 2**).

It should be noted that SubNcs hold the second position in the ranking of SubPc-type compounds after parent SubPc; their properties and optoelectronic applications deserve a separate circumstantial review.

5. Current status and perspectives

Work on design of D/A junctions with various halogen-substituted subphthalocyanines was initiated back in 2009 [64]. In the last few years it is Cl_6SubPc that has become one of the most efficient acceptors in both bilayer and blended heterojunctions. Many authors now use Cl_6SubPc as a reference when introducing their newly synthesized compounds belonging to the subphthalocyanine family in photovoltaic devices [16, 17, 24, 27, 28, 31–33, 66], as was commonly done earlier with C_{60}. However, the critical analysis of the current literature suggests that Cl_6SubPc still holds the lead among competitors. It combines availability (ease of synthesis and good yield), versatility of deposition (both wet and dry methods are available), appropriate color characteristics (position and intensity of the Q-band), and stability. Such benefits stimulate appearance of new heterojunctions designs employing this compound and steady interest in future research. For instance, such drawbacks as low charge carrier mobility in thin films can be overcome using morphology engineering [35]. Doping of the Cl_6SubPc molecular matrix with appropriate agents could be another option for improvement of the conducting properties [30]. Or vice versa, Cl_6SubPc can be doped into (mixed with) another acceptor to form a ternary blended junction [41].

Meanwhile, the search for new electron acceptors for photovoltaics among the subphthalocyanines with electron-withdrawing substituents on the periphery is underway. Two interesting approaches have been proposed last year by Torres and coworkers, who are the main newsmakers in this field. The first one consists in the synthesis of subphthalocyanines hexacyanated at the same peripheral positions as in Cl_6SubPc [31]. Unfortunately, due to the inherent instability of hexa-substituted derivatives, only slightly cyanated compounds were obtained and characterized, which have two cyano groups in only one isoindole unit, while the other two still bear two chlorines each, as in the parent Cl_6SubPc (**Figure 1**). Even for such CN_4Cl_2SubPc complex, there are indications on the increased mobility of charge carriers in vacuum-deposited films [31].

The second approach addresses the synthesis of new SubNc-type compounds with chlorines in the outer benzene rings [66]. Authors chose to directly use the dodeca-substituted derivative having four Cl atoms in each outer benzene ring, which unluckily is insoluble, whereas the entire work was targeted at making devices with a solution-processed BHJ[5] only. In this respect another work of Bender et al. [67] is worth mentioning, in which the authors argue that due to the nature of synthesis all of the SubNcs, both commercially available or obtained using the published protocols, in fact represent a mixture of derivatives randomly chlorinated at bay position. That is, in addition to the axial chlorine, part of the molecules inevitably contains chlorine atoms attached to the inner benzene rings of the naphthalene moieties. The outer benzene rings remain always hydrogenated. The presence of 1.13–2.96 chlorines per molecule on average was estimated [67]. Curiously, chlorination of all available bay positions in this compound would yield the Cl_6SubNc compound that has not been described yet.

The electron-accepting properties of the Cl_6SubPc molecule can be further enhanced by substituting the carbon atoms not bonded to chlorine with the more electronegative nitrogen. Such subporphyrazine-type compounds are synthesized in our group [24, 68]. However, the deep-lying HOMOs often cause a decrease in the specific conductivity of the bulk material. As with many other synthetic approaches, an exact balance must be maintained between the individual properties of a molecule and the photoelectrical properties (including morphological issues) of a solid.

Eventually, the group from Kyushu University developed the SubPc-type compounds, in which two chlorines in each benzene ring are (a) replaced by the -S-(C=O)-S- semicircle, or (b) peripheral benzenes in SubPc were directly replaced with the electron-withdrawing 1,3-dithiole-2-one units [69]. The deep bowl depths and curvatures of the formed SubPc and SubPz cores (*cf.* **Figure 1**) motivated authors to investigate the bimolecular concave-convex interactions with fullerenes in the co-crystals, as a first step to the fabrication of photovoltaically active materials.

6. Conclusion

Molecular properties of the Cl_6SubPc compound, such as redox potentials and positions of the absorption bands, fluorescence quantum yield, solubility, and stability, make it a material of choice when it comes to fabrication of a small-molecule based optoelectronic device, with almost any schematic. After a proper optimization of the donor material in the P/N junction (N = Cl_6SubPc) and with corresponding device composition, the open-circuit voltages above 1.3 V can be achieved. Further progress in the power conversion efficiency is limited mostly by the density of current leaking through the illuminated device, a parameter strongly dependent on the mutual arrangement of Cl_6SubPc molecules either in a layer (for PHJ) or in the interpenetrating network (if it exists) and on the morphology of the heterointerface at the nanoscale. The questions whether the electrical dipole or symmetry of the molecule could affect the generation of charge carriers by the junction do not seem to be of serious practical importance, but are very interesting for fundamental understanding of the photovoltaic process proper.

---

[5] The authors claimed that SubNcs have never been tested as either donors or acceptors in solution-processed BHJ solar cells [66]. This is not correct since in 2013, Yang and coworkers have published their data on both solution-processed and vacuum-evaporated BHJ with SubNc donor (acceptor was $PC_{70}BM$ or C_{70}, respectively) [39]. The optimized devices showed promising efficiency of 4.0 and 4.4% at room temperature (**Table 2**).

Acknowledgements

This work was supported by the Russian Science Foundation (grant #17-13-01522).

Conflict of interest

The authors declare no conflict of interest.

Author details

Georgy L. Pakhomov[1,2], Vlad V. Travkin[2] and Pavel A. Stuzhin[1*]

1 Ivanovo State University of Chemistry and Technology, Ivanovo, Russia

2 Institute for Physics of Microstructures of the Russian Academy of Sciences, Nizhny Novgorod, Russia

*Address all correspondence to: stuzhin@isuct.ru

IntechOpen

© 2019 The Author(s). Licensee IntechOpen. This chapter is distributed under the terms of the Creative Commons Attribution License (http://creativecommons.org/licenses/by/3.0), which permits unrestricted use, distribution, and reproduction in any medium, provided the original work is properly cited.

References

[1] Claessens CG, González-Rodríguez D, Rodríguez-Morgade MS, Medina A, Torres T. Subphthalocyanines, subporphyrazines, and subporphyrins: Singular nonplanar aromatic systems. Chemical Reviews. 2014;**114**:2192-2277. DOI: 10.1021/cr400088w

[2] Shimizu S, Kobayashi N. Structurally-modified subphthalocyanines: Molecular design towards realization of expected properties from the electronic structure and structural features of subphthalocyanine. Chemical Communications. 2014;**50**:6949-6966. DOI: 10.1039/C4CC01526F

[3] Rio Y, Rodriguez-Morgade MS, Torres T. Modulating the electronic properties of porphyrinoids: A voyage from the violet to the infrared regions of the electromagnetic spectrum. Organic & Biomolecular Chemistry. 2008;**6**:1877-1894. DOI: 10.1039/b800617b

[4] Morse GE, Bender TP. Boron subphthalocyanines as organic electronic materials. ACS Applied Materials & Interfaces. 2012;**4**:5055-5068. DOI: 10.1021/am3015197

[5] Grant TM, Josey DS, Sampson KL, Mudigonda T, Bender TP, Lessard BH. Boron subphthalocyanines and silicon phthalocyanines for use as active materials in organic photovoltaics. The Chemical Record. 2019;**19**:1-21. DOI: 10.1002/tcr.201800178

[6] Schwarze M, Tress W, Beyer B, Gao F, Scholz R, Poelking C, et al. Band structure engineering in organic semiconductors. Science. 2016;**352**:1446-1449. DOI: 10.1126/science.aaf0590

[7] Tang ML, Oh JH, Reichardt AD, Bao Z. Chlorination: A general route toward electron transport in organic semiconductors. The Journal of the American Chemical Society. 2009;**131**:3733-3740. DOI: 10.1021/ja809045s

[8] Shao JY, Cui BB, Tang JH, Zhong YW. Resistive memory switching of transition-metal complexes controlled by ligand design. Coordination Chemistry Reviews. 2019;**393**:21-36. DOI: 10.1016/j.ccr.2019.05.010

[9] Martynov AG, Safonova EA, Tsivadze AY, Gorbunova YG. Functional molecular switches involving tetrapyrrolic macrocycles. Coordination Chemistry Reviews. 2019;**387**:325-347. DOI: 10.1016/j.ccr.2019.02.004

[10] Hohnholz D, Steinbrecher S, Hanack M. Applications of phthalocyanines in organic light emitting devices. Journal of Molecular Structure. 2000;**521**:231-237. DOI: 10.1016/S0022-2860(99)00438-X

[11] Sakakibara Y, Bera RN, Mizutani T, Ishida K, Tokumoto M, Tani T. Photoluminescence properties of magnesium, chloroaluminum, bromoaluminum, and metal-free phthalocyanine solid films. The Journal of Physical Chemistry B. 2001;**105**:1547-1553. DOI: 10.1021/jp002943o

[12] Plint TG, Lessard BH, Bender TP. Doping chloro boron subnaphthalocyanines and chloro boron subphthalocyanine in simple OLED architectures yields warm white incandescent-like emissions. Optical Materials. 2018;**75**:710-718. DOI: 10.1016/j.optmat.2017.11.028

[13] Diaz DD, Bolink HJ, Cappelli L, Claessens CG, Coronado E, Torres T. Subphthalocyanines as narrow band red-light emitting materials. Tetrahedron Letters. 2007;**48**:4657-4660. DOI: 10.1016/j.tetlet.2007.05.036

[14] Melville OA, Lessard BH, Bender TP. Phthalocyanine-based organic thin-film transistors: A review of recent advances. ACS Applied Materials & Interfaces. 2015;**7**:13105-13118. DOI: 10.1021/acsami.5b01718

[15] Yasuda T, Tsutsui T. n-channel organic field-effect transistors based on boron-subphthalocyanine. Molecular Crystals and Liquid Crystals. 2007;**462**:3-9. DOI: 10.1080/15421400601009278

[16] Zhang J, Tan HS, Guo X, Facchetti A, Yan H. Material insights and challenges for non-fullerene organic solar cells based on small molecular acceptors. Nature Energy. 2018;**3**:720-731. DOI: 10.1038/s41560-018-0181-5

[17] Ebenhoch B, Prasetya NBA, Rotello VM, Cooke G, Samuel IDW. Solution-processed boron subphthalocyanine derivatives as acceptors for organic bulk-heterojunction solar cells. Journal of Materials Chemistry A. 2015;**3**:7345-7352. DOI: 10.1039/C5TA00715A

[18] Hancox I, New E, Jones TS. Utilising solution processed zirconium acetylacetonate as an electron extracting layer in both regular and inverted small molecule organic photovoltaic cells. Organic Electronics. 2015;**23**:105-109. DOI: 10.1016/j.orgel.2015.04.019

[19] Sullivan P, Schumann S, Da Campo R, Howells T, Duraud A, Shipman M, et al. Ultra-high voltage multijunction organic solar cells for low-power electronic applications. Advanced Energy Materials. 2013;**3**: 239-244. DOI: 10.1002/aenm.201200560

[20] Castrucci JS, Garner RK, Dang JD, Thibau E, Lu ZH, Bender TP. Characterization of μ-oxo-(BsubPc)2 in multiple organic photovoltaic device architectures: Comparing against and combining with Cl-BsubPc. ACS Applied Materials & Interfaces. 2016;**8**:24712-24721. DOI: 10.1021/acsami.6b06717

[21] Lee H, Ahn SW, Ryu SH, Ryu BK, Lee MH, Cho SW, et al. Interfacial electronic structure of Cl_6SubPc non-fullerene acceptors in organic photovoltaics using soft X-ray spectroscopies. Physical Chemistry Chemical Physics. 2017;**19**:31628-31633. DOI: 10.1039/C7CP04876A

[22] Sullivan P, Duraud A, Hancox I, Beaumont N, Mirri G, Tucker JHR, et al. Halogenated boron subphthalocyanines as light harvesting electron acceptors in organic photovoltaics. Advanced Energy Materials. 2011;**1**:352-355. DOI: 10.1002/aenm.201100036

[23] Beaumont N, Castrucci JS, Sullivan P, Morse GE, Paton AS, Lu ZH, et al. Acceptor properties of boron subphthalocyanines in fullerene free photovoltaics. The Journal of Physical Chemistry C. 2014;**118**:14813-14823. DOI: 10.1021/jp503578g

[24] Travkin V, Yunin P, Stuzhin P, Pakhomov G. Characterization of vacuum-deposited films of hexachloro-substituted subphthalocyanines for photovoltaic applications. Materials Today Proceedings. 2019. DOI: 10.1016/j.matpr.2019.07.708

[25] Verreet B, Cnops K, Cheyns D, Heremans P, Stesmans A, Zango G, et al. Decreased recombination through the use of a non-fullerene acceptor in a 6.4% efficient organic planar heterojunction solar cell. Advanced Energy Materials. 2014;**4**:1301413. DOI: 10.1002/aenm.201301413

[26] Sampson KL, Jiang X, Bukuroshi E, Dovijarski A, Raboui H, Bender TP, et al. A comprehensive scope of peripheral and axial substituent effect on the spectroelectrochemistry of boron subphthalocyanines. The Journal of Physical Chemistry A.

2018;**122**:4414-4424. DOI: 10.1021/acs.jpca.8b02023

[27] Duan C, Zango G, Iglesias MG, Colberts FJM, Wienk MM, Martinez-Diaz MV, et al. The role of the axial substituent in subphthalocyanine acceptors for bulk-heterojunction solar cells. Angewandte Chemie, International Edition. 2017;**56**:148-152. DOI: 10.1002/anie.201608644

[28] Castrucci JS, Josey DS, Thibau E, Lu ZH, Bender TP. Boron subphthalocyanines as triplet harvesting materials within organic photovoltaics. The Journal of Physical Chemistry Letters. 2015;**6**:3121-3125. DOI: 10.1021/acs.jpclett.5b01254

[29] Josey DS, Castrucci JS, Dang JD, Lessard BH, Bender TP. Evaluating thiophene electron-donor layers for the rapid assessment of boron subphthalocyanines as electron acceptors in organic photovoltaics: Solution or vacuum deposition? A European Journal of Chemical Physics and Physical Chemistry. 2015;**16**:1245-1250. DOI: 10.1002/cphc.201402751

[30] Cnops K, Zango G, Genoe J, Heremans P, Martinez-Diaz MV, Torres T, et al. Energy level tuning of non-fullerene acceptors in organic solar cells. Journal of the American Chemical Society. 2015;**137**:8991-8997. DOI: 10.1021/jacs.5b02808

[31] Zango G, Sakurai T, Urones B, Saeki H, Matsuda W, Martinez-Diaz MV, et al. Peripherally cyanated subphthalocyanines as potential n-type organic semiconductors. Chemistry–A European Journal. 2018;**24**: 8331-8342. DOI: 10.1002/chem.201801190

[32] Huang X, Hu M, Zhao X, Li C, Yuan Z, Liu X, et al. Subphthalocyanine triimides: Solution processable bowl-shaped acceptors for bulk heterojunction solar cells. Organic Letters. 2019;**21**:3382-3386. DOI: 10.1021/acs.orglett.9b01130

[33] Hang H, Wu X, Xu Q, Chen Y, Li H, Wang W, et al. Star-shaped small molecule acceptors with a subphthalocyanine core for solution-processed non-fullerene solar cells. Dyes and Pigments. 2019;**160**:243-251. DOI: 10.1016/j.dyepig.2018.07.050

[34] Hains AW, Liang Z, Woodhouse MA, Gregg BA. Molecular semiconductors in organic photovoltaic cells. Chemical Reviews. 2010;**110**:6689-6735. DOI: 10.1021/cr9002984

[35] Yunin PA, Travkin VV, Sachkov YI, Koptyaev AI, Stuzhin PA, Pakhomov GL. Increasing efficiency of hybrid p-CuI/n-Cl_6SubPc heterojunction through the interface engineering. Applied Surface Science. submitted

[36] Bredas JL. Mind the gap! Materials Horizons. 2014;**1**:17-19. DOI: 10.1039/c3mh00098b

[37] Morse GE, Paton AS, Lough A, Bender TP. Chloro boron subphthalocyanine and its derivatives: Dyes, pigments or somewhere in between? Dalton Transactions. 2010;**39**:3915-3922. DOI: 10.1039/b922199a

[38] Ma B, Miyamoto Y, Woo CH, Fréchet JMJ, Zhang F, Liu Y. Solution processable boron subphthalocyanine derivatives as active materials for organic photovoltaics. In: Proceedings of the SPIE; 7416 Organic Photovoltaics X. 2009. p. 74161E. DOI: 10.1117/12.825372

[39] Chen G, Sasabe H, Sano T, Wang X-F, Hong Z, Kido J, et al. Chloroboron (III) subnaphthalocyanine as an electron donor in bulk heterojunction photovoltaic cells. Nanotechnology. 2013;**24**:484007. DOI: 10.1088/0957-4484/24/48/484007

[40] Verreet B, Müller R, Rand BP, Vasseur K, Heremans P. Structural templating of chloro-aluminum phthalocyanine layers for planar and bulk heterojunction organic solar cells. Organic Electronics. 2011;**12**:2131-2139. DOI: 10.1016/j.orgel.2011.08.031

[41] Trinh C, Kirlikovali KO, Bartynski AN, Tassone CJ, Toney MF, Burkhard GF, et al. Efficient energy sensitization of C_{60} and application to organic photovoltaics. Journal of the American Chemical Society. 2013;**135**:11920-11928. DOI: 10.1021/ja4043356

[42] Cnops K, Rand BP, Cheyns D, Verreet B, Empl MA, Heremans P. 8.4% efficient fullerene-free organic solar cells exploiting long-range exciton energy transfer. Nature Communications. 2014;**5**:3406. DOI: 10.1038/ncomms4406

[43] Wild M, Berner S, Suzuki H, Yanagi H, Schlettwein D, Ivan S, et al. A novel route to molecular self-assembly: Self-intermixed monolayer phases. A European Journal of Chemical Physics and Physical Chemistry. 2002;**3**:881-885. DOI: 10.1002/1439-7641(20021018)3:10<881:AID-CPHC881>3.0.CO;2-P

[44] Pandey R, Gunawan AA, Mkhoyan KA, Holmes RJ. Efficient organic photovoltaic cells based on nanocrystalline mixtures of boron subphthalocyanine chloride and C_{60}. Advanced Functional Materials. 2012;**22**:617-624. DOI: 10.1002/adfm.201101948

[45] Huang Y, Kramer EJ, Heeger AJ, Bazan GC. Bulk heterojunction solar cells: morphology and performance relationships. Chemical Reviews. 2014;**114**:7006-7043. DOI: 10.1021/cr400353v

[46] Drozdov MN, Drozdov YN, Pakhomov GL, Travkin VV, Yunin PA, Razumov VF. Depth profiling of fullerene-containing structures by time-of-flight secondary ion mass spectrometry. Technical Physics Letters. 2013;**39**:1097-1100. DOI: 10.1134/S1063785013120183

[47] Chen WX, Xu ZD, Li WZ. Photoconductivity of C_{60}-doped phthalocyanine composites. Journal of Photochemistry and Photobiology, A: Chemistry. 1995;**88**:179-182. DOI: 10.1016/1010-6030(94)03999-B

[48] Chen ZH, Xie YM, Li J, Zhan MX. Studies on the formation of charge transfer complex between fullerene and zinc phthalocyanine. Chemical Journal of Chinese Universities. 1997;**18**:1534-1536. Available from: http://www.cjcu.jlu.edu.cn/EN/Y1997/V18/I9/1534

[49] Ruani G, Dediu V, Liess M, Lunedei E, Michel R, Muccini M, et al. Photoinduced charge transfer in complex architectured films of C_{60} and donor-like molecules. Synthetic Metals. 1999;**103**:2392-2394. DOI: 10.1016/S0379-6779(98)00297-5

[50] Ruani G, Fontanini C, Murgia M, Taliani C. Weak intrinsic charge transfer complexes: A new route for developing wide spectrum organic photovoltaic cells. The Journal of Chemical Physics. 2002;**116**:1713-1719. DOI: 10.1063/1.1429235

[51] Maennig B, Drechsel J, Gebeyehu D, Simon P, Kozlowski F, Werner A, et al. Organic p-i-n solar cells. Applied Physics A. 2004;**79**:1-14. DOI: 10.1007/s00339-003-2494-9

[52] Meiss J, Merten A, Hein M, Schuenemann C, Schäfer S, Tietze M, et al. Fluorinated zinc phthalocyanine as donor for efficient vacuum-deposited organic solar cells. Advanced Functional Materials. 2012;**22**:405-414. DOI: 10.1002/adfm.201101799

[53] Pfuetzner S, Mickel C, Jankowski J, Hein M, Meiss J, Schuenemann C, et al. The influence of substrate heating on morphology and layer growth in C_{60}:ZnPc bulk heterojunction solar cells. Organic Electronics. 2011;**12**:435-441. DOI: 10.1016/j.orgel.2010.12.007

[54] Gilchrist JB, Basey-Fisher TH, Chang S, Scheltens F, McComb DW, Heutz S. Uncovering buried structure and interfaces in molecular photovoltaics. Advanced Functional Materials. 2014;**24**:6473-6483. DOI: 10.1002/adfm.201400345

[55] Simon P, Maennig B, Lichte H. Conventional electron microscopy and electron holography of organic solar cells. Advanced Functional Materials. 2004;**14**:669-676. DOI: 10.1002/adfm.200304498

[56] Ferro VR, Garcia VJM, Claessens CG, Poveda LA, Gonzalez-Jonte RH. The axial coordination in subphthalocyanines. Geometrical and electronic aspects. Journal of Porphyrins and Phthalocyanines. 2001;**5**:491-499. DOI: 10.1002/jpp.338

[57] Ferro VR, Poveda LA, Gonzalez-Jonte RH, VJM G, Torres T, del Rey B. Molecular electronic structure of subphthalocyanine macrocycles. Journal of Porphyrins and Phthalocyanines. 2000;**4**:611-620. DOI: 10.1002/1099-1409(200009/10)4:6<611::AID-JPP230>3.0.CO;2-A

[58] Waters MJ, Hashemi D, Shi G, Kioupakis E, Kieffer J. Predictive simulations for tuning electronic and optical properties of SubPc derivatives. Journal of Electronic Materials. 2019;**48**:2962-2970. DOI: 10.1007/s11664-019-06961-w

[59] Morris SE, Bilby D, Sykes ME, Hashemi H, Waters MJ, Kieffer J, et al. Effect of axial halogen substitution on the performance of subphthalocyanine based organic photovoltaic cells. Organic Electronics. 2014;**15**:3660-3665. DOI: 10.1016/j.orgel.2014.09.048

[60] Isaacs EB, Sharifzadeh S, Ma B, Neaton JB. Relating trends in first-principles electronic structure and open-circuit voltage in organic photovoltaics. The Journal of Physical Chemistry Letters. 2011;**2**:2531-2537. DOI: 10.1021/jz201148k

[61] Linares M, Beljonne D, Cornil J, Lancaster K, Bredas JL, Verlaak S, et al. On the interface dipole at the pentacene-fullerene heterojunction: A theoretical study. The Journal of Physical Chemistry C. 2010;**114**:3215-3224. DOI: 10.1021/jp910005g

[62] Breuer T, Karthäuser A, Witte G.Effects of molecular orientation in acceptor–donor interfaces between pentacene and C_{60} and diels–alder adduct formation at the molecular interface. Advanced Materials Interfaces. 2016;**3**:1500452. DOI: 10.1002/admi.201500452

[63] Rand BP, Richter H, editors. Organic Solar Cells: Fundamental, Devices and Upscaling. 1st ed. Boca Raton: CRC press, Taylor & Francis Group; 2014. 810p. DOI: 10.1201/b17301

[64] Gommans H, Aernouts T, Verreet B, Heremans P, Medina A, Claessens CG, et al. Perfluorinated subphthalocyanine as a new acceptor material in a small-molecule bilayer organic solar cell. Advanced Functional Materials. 2009;**19**:3435-3439. DOI: 10.1002/adfm.200900524

[65] Sai N, Gearba R, Dolocan A, Tritsch JR, Chan WL, Chelikowsky JR, et al. Understanding the interface dipole of copper phthalocyanine (CuPc)/C_{60}: Theory and experiment. The Journal of Physical Chemistry Letters. 2012;**3**:2173-2177. DOI: 10.1021/jz300744r

[66] Duan C, Guzmán D, Colberts FJM, Janssen RAJ, Torres T.

Subnaphthalocyanines as electron acceptors in polymer solar cells: Improving device performance by modifying peripheral and axial substituents. Chemistry–A European Journal. 2018;**24**:6339-6343. DOI: 10.1002/chem.201800596

[67] Dang JD, Josey DS, Lough AJ, Li Y, Sifate A, Luc ZH, et al. The mixed alloyed chemical composition of chloro-(chloro)n-boron subnaphthalocyanines dictates their physical properties and performance in organic photovoltaic devices. Journal of Materials Chemistry A. 2016;**4**:9566-9577. DOI: 10.1039/C6TA02457B

[68] Stuzhin PA, Skvortsov IA, Zhabanov YA, Somov NV, Razgonyaev OV, Nikitin IA, et al. Subphthalocyanine azaanalogues–Boron(III) subporphyrazines with fused pyrazine fragments. Dyes and Pigments. 2019;**162**:888-897. DOI: 10.1016/j.dyepig.2018.11.006

[69] Wang Y, Uchihara K, Mori S, Furuta H, Shimizu S. 1,3-Dithiole-2-one-fused subphthalocyanine and subporphyrazine: Synthesis and properties arising from the 1,3-Dithiole-2-one units. Organic Letters. 2019;**21**:3103-3107. DOI: 10.1021/acs.orglett.9b00752

Chapter 6

Peculiarities of Refractory Borides Formation during Mechanical Alloying IV-V Group Transition Metals with Boron in Planetary Mill

Maria P. Savyak and Alex B. Melnick

Abstract

Mechanical alloying in the transition IV-V group metal-boron systems runs by the two following mechanisms: mechanically induced reaction of self-propagating synthesis determined by the enthalpy of refractory compound formation and capability to form substitutional solid solution through replacement of a metal atom by boron atoms; and diffusion-controlled process when a supersaturated interstitial solid solution prevails and its *bcc* lattice gradually transforms to the hexagonal lattice of the MeB_2 phase at a critical boron content. The domination of one of the above mechanisms is determined by capability of boron to form substitutional or interstitial solid solution. In the case of formation of combined (SSS and ISS) solid solutions, domination of a mechanism is determined by the interatomic bond strength as well as by the intensity of mechanical alloying. The method for calculation of the free Gibbs energy of the interstitial and substitutional solid solutions on the basis of the regular solution model was developed. It was shown that during milling tantalum and boron in a planetary mill, at first the formation of a combined solid solution occurs where two boron atoms replace one tantalum atom. Both the mechanisms of solid solution formation decrease the solution Gibbs energy. When a SSS dominates over the formation of an ISS, the Gibbs energy acquires a minimum value at a concentration of boron in tantalum of 50 at%, which leads to the solution decomposition.

Keywords: transition metals, refractory borides, Gibbs energy, solid solution, mechanical alloying, modeling

1. Introduction

Refractory metal borides have a high melting point, high thermal and electrical conductivity, low linear coefficient of thermal expansion, excellent corrosion resistance, and very high microhardness. Therefore, they attract much attention as promising materials for application. The potential application of such materials could be jet engine parts, armor plates, cutting tools, dies, etc. [1]. A large number of studies are devoted to these materials and in particular to methods for their preparation. As a rule, borides are obtained at high temperature and, therefore, the grain size of the formed boride is big. However, high enthalpy of borides points that these compounds can be obtained at reduced temperature (for example, formation enthalpy ΔH (in kJ/mol) is 266 for TiB_2, 333 for HfB_2, 182–191 for TaB, 181–192 for TaB_2, 221 for NbB, 174 for NbB_2, 130 for VB, and 259 for VB_2 [2]). This can be performed with the help of a planetary mill where extremely high centrifugal forces appear under milling due to the disk and the vials rotation in the counter direction. Milling process of a solid state powder, where the powder particles are subjected to high energetic impacts by the balls in a vial, which involves repeated cold welding, fracturing, and rewelding of powder particles, is called mechanical alloying [3]. Numerous studies have been devoted to production of transition metal borides using a mechanical alloying [4–11]. This method allows one to obtain a nanocrystalline structure of these compounds, which helps to reduce the sintering temperature of borides to achieve a dense state of material. Ball milling can induce self-propagating reaction in such highly exothermic powder mixtures as boron and a transition metal. This process is called the self-propagating mechanically induced synthesis (SMS) [4]. SMS is characterized by ignition temperature T_i, at which the self-propagating reaction begins, and temperature T_m, to which the powder is heated between the ball collisions. The latter increases at the expense of the kinetic energy of the balls. As established experimentally, such an increase takes place up to 350°C, which is not enough to initiate the SMS process. In the standard SMS process $T_i > T_m$ and, hence, no ignition occurs. However, T_i decreases during the milling process under the action of several factors: the particles are refined, chemically active defects emerge, new dislocations and interfaces show up, etc. So when $T_i < T_m$, ignition does occur. Therefore the question arises: what initiates the SMS process in mechanical alloying of borides? One of the reasons is stresses arising in the metal due to various defects of dislocations and solid solutions induced in the metal during milling. Therefore, it is necessary to carefully study the structural changes in the metal after milling. The peculiarities of structure transformation under milling transition metals with boron have not clarified yet. The mechanisms of SMS during formation of the transition metal borides by mechanical alloying are still under discussion as well. It has been established that in the course of mechanical processing of Nb + B mixtures, formation of boron-in-niobium solid solution takes place followed by gradual precipitation of the NbB_2 phase [8]. However, questions concerning the influence of the structure and atom size on the peculiarities of boride formation during mechanical alloying have been scarcely considered so far. The transition metals of group IV Ti, Zr, and Hf have a *hexagonal close-packed* (*hcp*) structure while the transition metals of group V Nb and Ta have a body-centered (*bcc*) structure, which is characterized by a bigger free volume (space fill factors are 0.74 and 0.68, respectively). The group VI metals Cr, Mo, and W have a *bcc* structure as well, but their atoms are much smaller compared to group V metals and possess a stronger interatomic bond, which is confirmed by far higher Young modulus.

We selected the Ta-B system with different tantalum-to-boron ratios as a model to study the peculiarities of structure transformation under milling transition group V metals (*bcc* structure) with boron and Ti-B system as a model to study the peculiarities of structure transformation in the *hcp* structure.

The aim of the work was to study what phases can form during mechanical alloying of transition metals with different crystalline structures with boron in a planetary mill and how much the phase formation depends on the metal structure, the metal/boron atom size relationship, and the peculiarities of mechanical alloying process and to develop a model of solid solutions formation during milling using the Ta-B system as an example.

2. Experimental procedure

High purity tantalum and niobium powders with 10–60 μm particles were used. They were produced at the "Silmet" metallurgical plant by electrolytic reduction with hydrogenation and dehydrogenation for further purification. Titanium powder was produced at the Zaporizhzhya titanium-magnesium plant with a specific surface area of 0.1 m^2/g brand TG-T_B (−2 + 1) mm, zirconium (Dneprodzerzhinsky PA "PCP") was obtained by the calcium thermal method, vanadium by thermal dissociation of vanadium iodide. Black amorphous boron powder (B-99 grade, as per 1-92-15490) with a specific surface area of 11.2 m^2/g was used. The brand of such black amorphous boron powder was obtained by gaseous boron halogenide reduction with hydrogen. The metal-to-boron atoms ratio depended on a specific metal-boron compound to be produced: MeB (1Me:1B), and MeB_2 (1Me:2B). Milling was performed in an argon medium in a planetary mill AIR 015 M, which provides an acceleration of 45 g at an rotation speeds of the disk and vials of 735 and 1840 rev/min, respectively. The balls-to-powder mass ratio was 20:1. XRD analysis was carried out on an installation DRON3 under copper K*α* radiation. Broadening analysis of X-ray reflections estimated the coherent scattering domain (CSD) and the crystal lattice microdistortion ε. The microstructure was studied using a transmission electron microscopy (TEM) JEM-100CX. The steel balls were pre-milled with boron and tantalum at the same time as the process of milling was 50 min. Therefore, contamination of steel material was minimal. In the case of milling within 10 min, the contamination of iron was 0.4 mass%. The iron worn off from the milling balls was removed with HCl leaching under the effect of ultrasonic stirring.

3. Experimental results

3.1 Ti-B

Using titanium as an example, let us consider how borides are formed during milling *hcp* metals of group IV with boron. **Figure 1** demonstrates changes in XRD patterns of Ti-B (1:2) mixture depending on the milling time. The change in the crystal lattice volume of titanium after milling with boron, calculated by the Rietveld method, is shown in **Figure 2.** As seen, the lattice volume of titanium noticeably decreases after 5 min milling. Under milling, titanium undergoes significant plastic deformation, which is confirmed by the texture in the plane of the easiest slip (002) in **Figure 1b**. Boride phase nucleation occurs after 6 min of milling, therefore the lattice volume of titanium increases (**Figure 2**). Titanium completely transforms into titanium diboride TiB_2 after 7 min milling (**Figure 1d**). The electron microscopic studies showed that milling of powder mixtures Ti-B for 6 min leads to extraction of nanodisperse allocation no larger than 3–5 nm in the shells of particles. TEM microphotographs of obtained TiB_2 are shown in **Figure 3**. Flat, greatly loosened along the edges polycrystalline particles are the main components after 7 min milling. These particles consist of nanodisperse grains of titanium diboride. The main range of the grain sizes is 3–10 nm; however, there are also areas where they reach 15–20 nm. To estimate the effect of boron on the change in the crystal lattice volume of titanium after milling, we milled pure titanium powder in argon. We did not observe any formation of texture during milling of pure titanium; also, we did not reveal any changes in the size of the titanium crystalline lattice after 5 and 6 min milling.

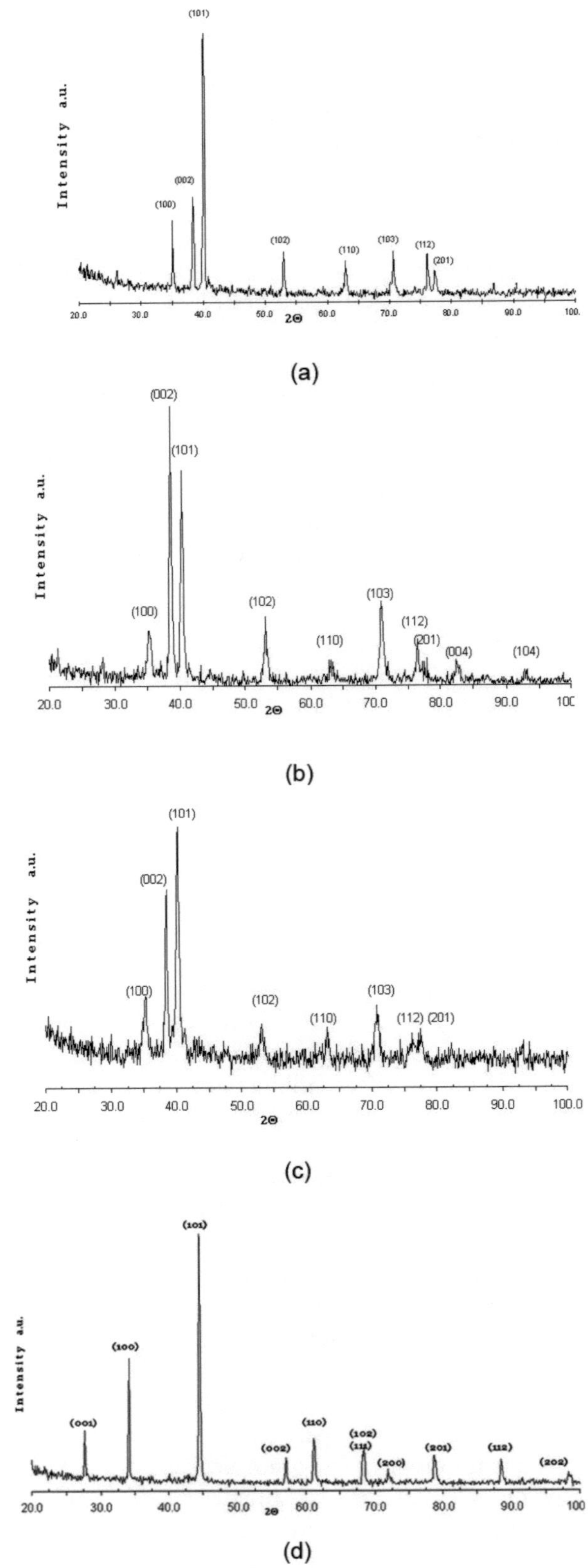

Figure 1.
XRD patterns of Ti-B mixture in the initial state (a), upon milling for (b) 5, (c) 6, and (d) 7 min (the indices a, b, c denote Ti, and d TiB_2).

3.2 Zr-B

Zirconium like titanium has the *hcp* structure. Zirconium diboride is formed in the same way as TiB_2. XRD patterns of ZrB_2 formed upon 6 min milling of Zr-B mixture are shown in **Figure 4**.

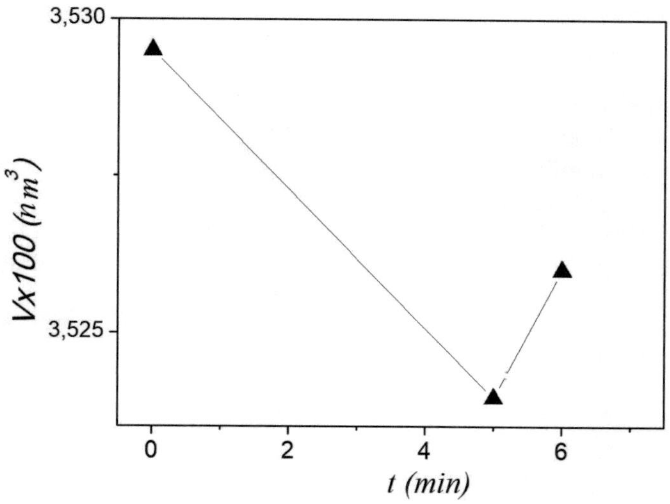

Figure 2.
Change in the crystal lattice volume of titanium after milling with boron calculated by the full-profile Rietveld method [9].

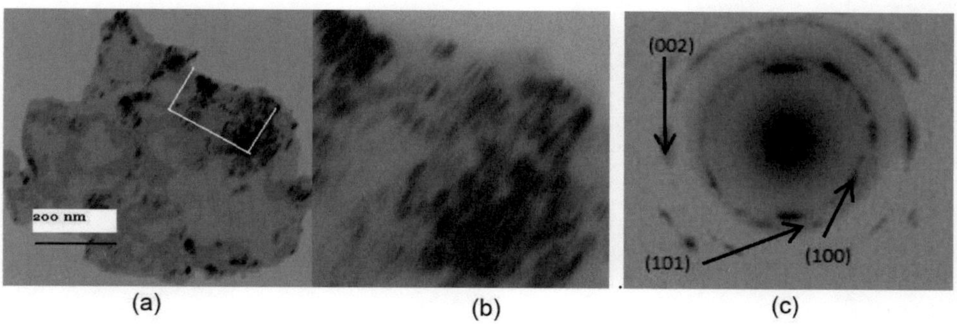

Figure 3.
TEM microphotograph of TiB_2 obtained upon milling Ti and B powders: (b) increased area accented in (a); (c) microelectron diffraction pattern.

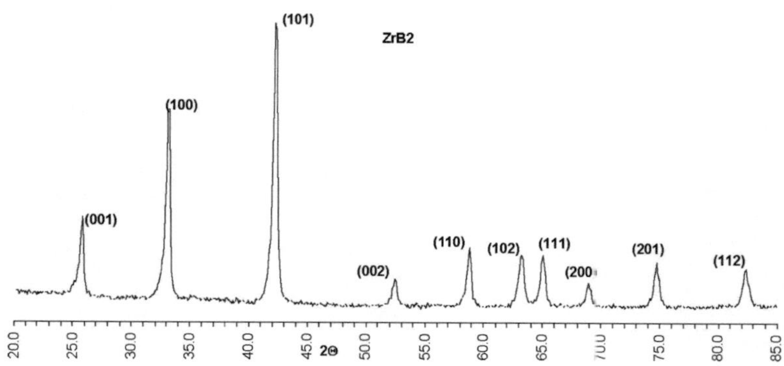

Figure 4.
XRD patterns of ZrB2 formed upon 6 min milling of Zr-B powders.

3.2.1 Milling of bcc *group V metals with boron*

A peculiarity of the group V transitions metals is that they have a *bcc* structure, in which there is much more free volume in comparison with the close-packed structures.

3.3 V-B

Figure 5 shows changes in the XRD patterns of V-B mixtures depending on the milling time at V:B ratios of 1:1 and 1:2. The diffraction lines shift toward smaller angles within 5 min milling, which indicates an increase in the lattice parameters at the both atomic ratios of the components. The vanadium lattice parameter (a = 0.3028 nm) for V:B = 1:1 after milling with boron for 5 min increases and is equal to 0.3034 nm. Within 20 min milling, the V:B = 1:1 mixture transforms into

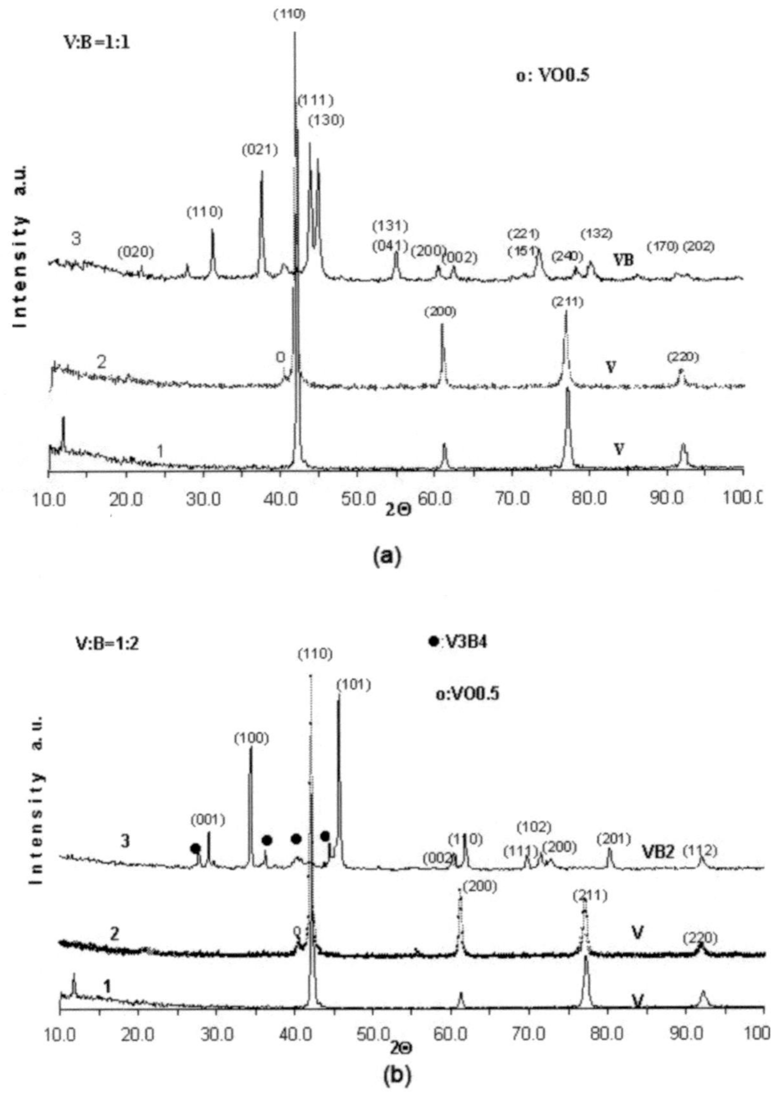

Figure 5.
XRD patterns of V-B powders upon milling for (a) (1) 0, (2) 5, (3) 20 min, (b) (1) 0, (2) 5, (3) 15 min [the indices on lines 2 denote V, on lines 3: (a) VB, (b) VB2].

the VB phase extremely rapidly, whereas the V:B = 1:2 mixture does into the VB_2 phase upon 15 min milling.

3.4 Nb-B

XRD patterns for the transformation of niobium into the borides NbB and NbB_2 under milling Nb-B mixtures are shown in **Figure 6**. Unlike vanadium characterized by increasing the lattice parameter within 5 min milling, in this case, at the ratio Nb:B = 1:1, the lattice parameter does not change, whereas at the ratio Nb:B = 1:2, it reduces and is equal to 0.3293 nm (a = 0.3300 nm for Nb), which indicates the formation of a substitutional solid solution. After niobium transformation into borides, in the XRD patterns, a strong niobium line (110) appears, shifted toward smaller angles, that is, the lattice parameter increases. This remaining niobium with an increased lattice parameter is evidence to the existence of parallel mechanisms of formation of boron-in-niobium solid solutions: on the one hand, a SSS is formed,

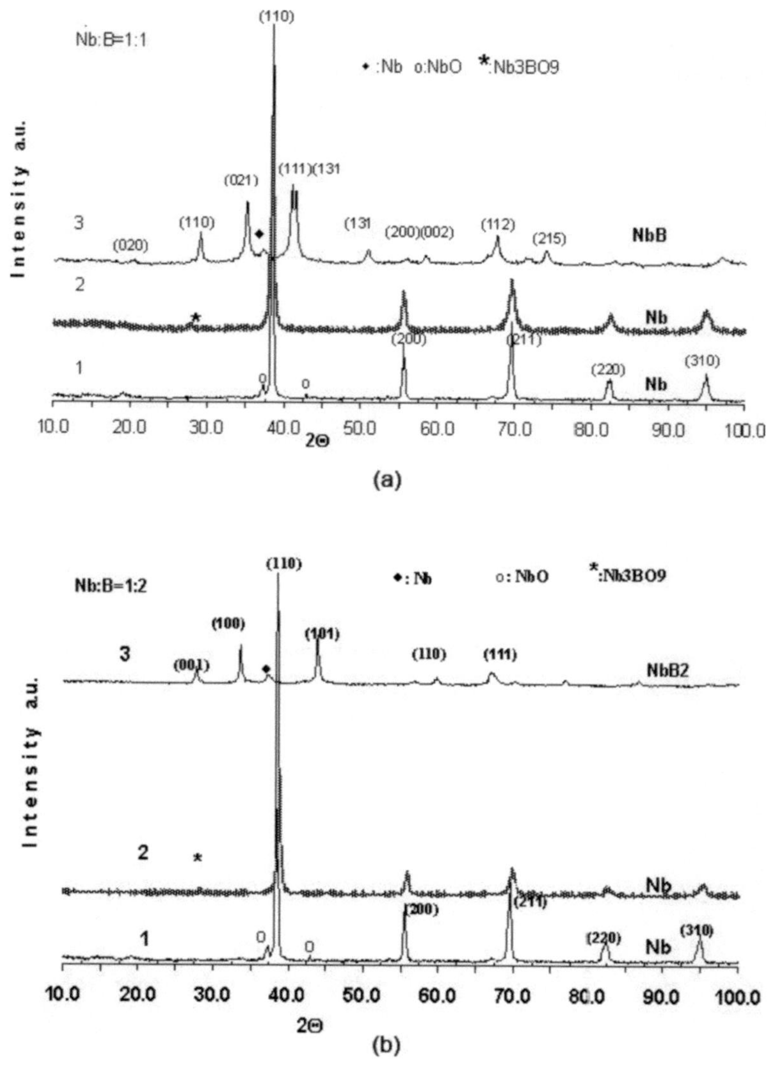

Figure 6.
XRD patterns of Nb-B mixture upon milling for (a) (1) 0, (2) 5, (3) 20 min, (b) (1) 0, (2) 5 and (3) 15 min [the indices on lines 1 denote Nb, on lines 3: (a) NbB, (b) NbB2].

which leads to a decrease in the lattice parameter; on the other hand, an ISS is formed, which leads to the lattice parameter increasing.

3.5 Ta-B

3.5.1 Powder mixture at the atomic ratio Ta:B = 1:1

In the case of 5 min milling of the Ta:B = 1:1 mixture, the resulted structure is defective, which is confirmed by weakening and broadening X-ray lines with remaining their positions (**Figure 7**). After 8 min milling, the tantalum lines positions are still the same (**Figure 8**). The 10 min milling transforms the tantalum powder into TaB with residual tantalum, whose line (110) is shifted toward smaller angles (**Figure 9**), which indicates that the lattice parameter increases owing to the formation of an interstitial boron-in-tantalum solid solution under milling. The facts that the tantalum lattice parameter does not change after milling for 5 and 10 min and that there is residual tantalum with increased lattice parameter after 10 min milling may be indicative of running two processes in parallel which differently influence the lattice parameter. Therefore, it can be assumed that both interstitial and substitutional solid solutions of boron in tantalum can be formed under milling. The lattice parameter in the Ta:B = 1:1 mixture increases for ISS and decreases for SSS. The powder mixtures containing 50 at% B show no change in the lattice parameter.

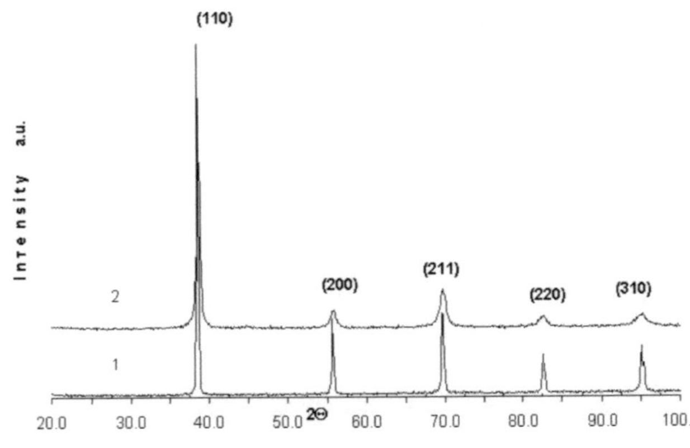

Figure 7.
XRD patterns of Ta:B = 1:1 powder mixture: before milling (1) upon milling for 5 min (2) (the indices in **Figures** *7 and* **8** *denote Ta).*

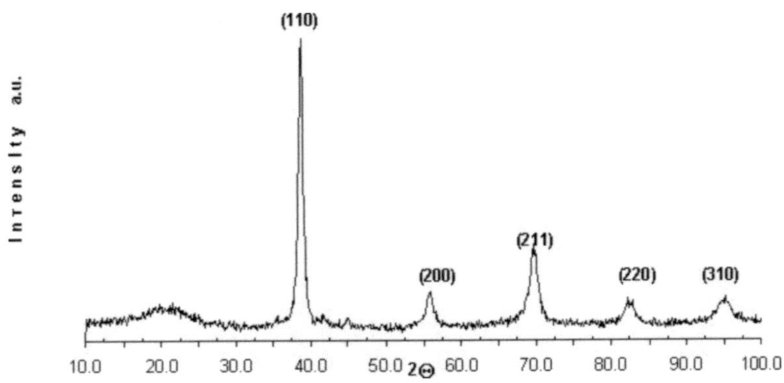

Figure 8.
XRD patterns of Ta:B = 1:1 powder mixture upon 8 min milling.

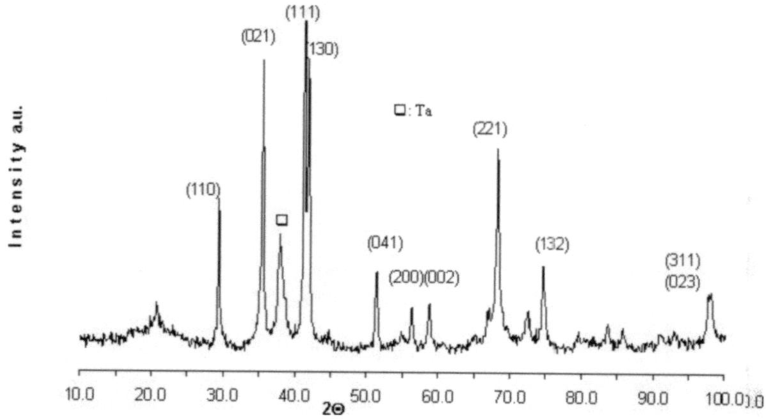

Figure 9.
XRD patterns of Ta:B = 1:1 powder mixture upon milling for 10 min (the indices denote TaB).

3.5.2 *Powder mixture at the atomic ratio Ta:B = 1:2*

Figure 10 compares XRD patterns recorded from powders milled during different periods of time. As low-intensity peaks for the mixture milled for 15 min are not seen in pattern 3 (**Figure 10a**) because of the scale factor, this pattern is enlarged in **Figure 10b**, where one can see the peaks corresponding to the TaB_2 phase. After milling for 5 min, the lines shift toward smaller angles and reveal that the lattice parameter changes from a = 0.3302 nm to a = 0.3315 nm, which indicates the fact of formation of an ISS. The intensive milling of the Ta:B = 1:2 powders for 5 min leads, like for Ta:B = 1:1 mixtures, to tantalum lattice distortion, which is confirmed by the broadening the X-ray lines and a lower intensity of reflection peaks under the impact of shock loads and shear deformation in the milling process. For the Ta: B = 1:2 mixture milled for 5 min, coherent scattering domain (CSD) size D is 28.91 nm and distortion ε is 0.004578. **Figure 11** shows XRD patterns from the Ta: B = 1:2 mixture milled for 15, 30, and 50 min. Milling for 15 min leads to an increase in *bcc* lattice distortion ε to 0.02394 with no significant change in the CSD size. The tantalum diboride lines appear as well. Therefore, the solvent metal lattice undergoes transformation (so-called concentration polymorphism) during the formation of a supersaturated solid solution. When milling time increases to 30 and 50 min, the tantalum diboride lines become more intensive. The results of performed calculations are the following: the lattice parameter a of the initial tantalum powder is 0.3303 nm and increases to 0.3314 nm after milling for 5 min, to 0.3364 nm for 15 min, and to 0.3368 nm for 50 min.

As established in [12], the change in the lattice parameter Δa to 0.0018 nm corresponds to the dissolution of 10 at% B. In this work, Δa is 0.0011 nm after 5 min milling, which corresponds to 7 at% B. The intensive milling of Ta and B powders can induce significant lattice distortion, which promotes diffusion of boron atoms into tetrahedral pores in the tantalum *bcc* lattice with the formation of an ISS, the decomposition of which leads to the appearance of hexagonal tantalum diboride (AlB_2 type). The diboride formed after 50 min milling is characterized by the following lattice parameters: a = 0.3086 nm and c = 0.3254 nm. The tantalum diboride has CSD D = 3.73 nm and distortion ε = 0.007851. In view of the above, one can conclude that occupation of tetrahedral pores in the tantalum *bcc* lattice becomes more preferable when boron concentration increases to 66 at%, which breaks the equilibrium between the SSS and ISS existing at 33 and 50 at% B. The lattice parameter increases after this mixture is milled for 5 min.

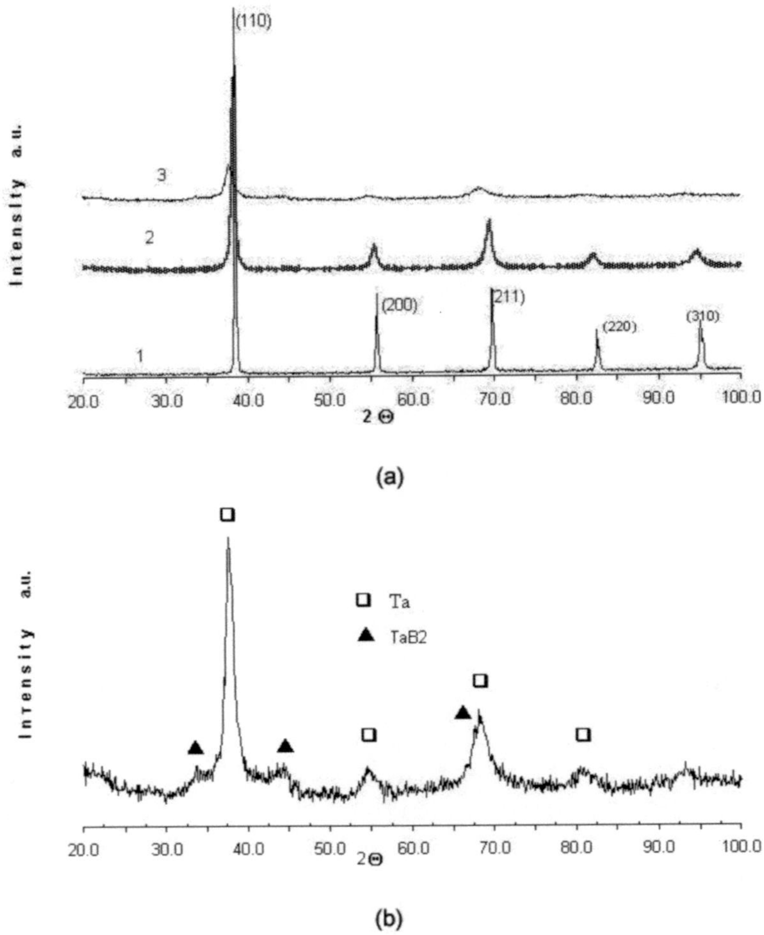

Figure 10.
XRD patterns of Ta:B = 1:2 powders in the initial state (1), upon milling for 5 (2), and 15 (3) min (the indices in a denote Ta). (b) Enlarged scale of above pattern 3 in (a).

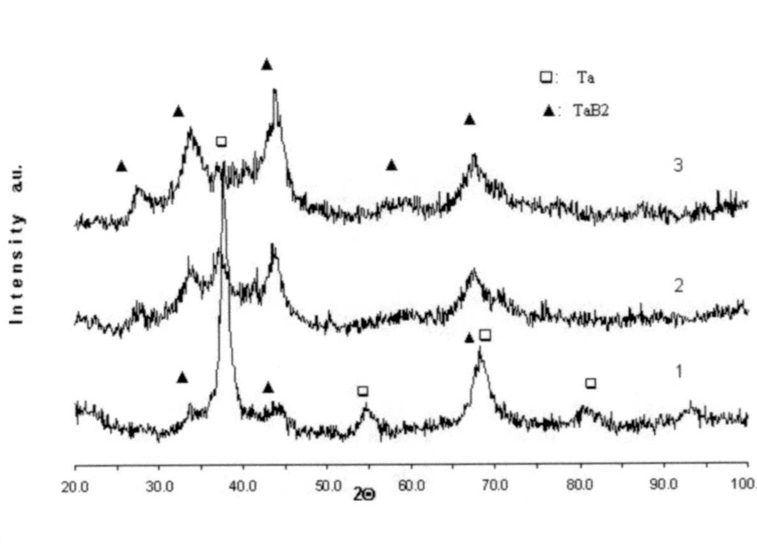

Figure 11.
XRD patterns of Ta:B = 1:2 powders mixture upon milling for 15 (1), 30 (2), 50 (3) min.

3.6 Modeling for process of boron-in-tantalum solid solution formation

In order to produce stable solid solutions, the authors first estimated stability of boron-in-tantalum ISS and SSS through determination of their Gibbs energy with taking into account that the latter is formed by the elastic energy (owing to the distortions in solid solutions) and the enthalpy of the system milling. Then, the solid solutions were studied in the frame of the regular solution model [13], using the following formula for the enthalpy of mixing n component alloys [14]:

$$\Delta H_{mix} = \sum_{\substack{i,j=1 \\ i \neq j}}^{n} c_i c_j \Omega_{ij}, \tag{1}$$

where Ω_{ij} is a parameter characterizing the interaction between i and j elements in a regular solution; $\Omega_{ij} = 4\Delta H^{ij}_{mix}$, c_i is the atomic fraction of the ith element; ΔH^{ij}_{mix} is the enthalpy of mixing for binary equiatomic alloys.

According to the Boltzmann hypothesis, the entropy of mixing n elements in a regular solution can be expressed as follows:

$$\Delta S_{mix} = -k_B \sum_{i=1}^{n} c_i \ln(c_i), \tag{2}$$

where c_i is the atomic fraction of the ith element, and k_B is the Boltzmann constant.

The solid solutions are crystals with a distorted lattice because their atoms have different sizes. Elastic distortions, arising from size discrepancy, can affect the free energy of the alloy as well. It is important to take into consideration different factors that contribute to the total enthalpy, such as the elastic energy ΔH_{el} and mixing enthalpy ΔH_{mix}:

$$\Delta H = \Delta H_{mix} + \Delta H_{el}, \tag{3}$$

We consider that the atomic volumes and local bulk moduli for solid solutions correspond to those for single-component systems. Since the components have different sizes, the lattice becomes distorted. Taking $\overline{V}$ as an average atomic volume of the solid solution, we will write equations for distortion of different atoms: $\varepsilon_1 = (V_{Ta} - \overline{V})/\overline{V}$ is the tantalum atom distortion in the absence of interstitial atoms nearby, $\varepsilon_2 = (2V_B - \overline{V})/\overline{V}$ is the distortion of two boron atoms when they replacing one tantalum atom in the *bcc* lattice in the absence of interstitial atoms nearby and V_{Ta} and V_B are the atomic volumes of tantalum and boron, respectively. Since two boron atoms and one replaced tantalum atom are commensurable in volume, it can be assumed that boron causes minimum distortion. It can penetrate into octahedral and tetrahedral pores of the *bcc* lattice. Then distortion of the interstitial boron atom will be $\varepsilon_3 = (V_B - Vp)/Vp$ (V_p is the pore volume). Finally, distortion of the tantalum and boron atoms that replace them in the presence of m interstitial atoms nearby will become $\varepsilon_4(m) = (V_{Ta} - (\overline{V} - m\alpha Vp))/(\overline{V} - m\alpha Vp)$ and $\varepsilon_5(m) = (2V_B - (\overline{V} - m\alpha Vp))/(\overline{V} - m\alpha Vp)$, respectively.

Here α is the parameter that characterizes the pore volume fraction corresponding to one tantalum atom. We take into account that $V_p = \frac{2}{3}\overline{V}$ and

α = 1/6 in case of an octahedral pore and $V_p = \frac{1}{6}\overline{V}$ and α = 1/4 in case of a tetrahedral pore. If the lattice is stable, its stresses are compensated and the following equation is valid:

$$\begin{aligned} &c_1 B_{Ta}\varepsilon_4(m-1)\left(\overline{V}-(m-1)\alpha V_P\right)+c_2 B_B\varepsilon_2\overline{V}+c_3 B_B\varepsilon_5(m-1)\left(\overline{V}-(m-1)\alpha V_P\right)\\ &\quad + c_4 B_{Ta}\varepsilon_4(m)\left(\overline{V}-m\alpha V_P\right)+c_5 B_B\varepsilon_5(m)\left(\overline{V}-m\alpha V_P\right)=0 \end{aligned} \tag{4}$$

where B_{Ta} and B_B are the bulk moduli for tantalum and boron. If c_{Ta} is the atomic fraction of tantalum in the solution and c_{Bs} and c_{Bi} are the atomic fractions of boron in substitutional and interstitial states, then the concentration coefficients for SSS and ISS with boron in octahedral pores will be as follows:

$$c_1 = c_{Ta}\left(1-\frac{6c_{Bi}}{c_{Ta}+\frac{c_{Bs}}{2}}+INT\left[\frac{6c_{Bi}}{c_{Ta}+\frac{c_{Bs}}{2}}\right]\right);$$

$$c_2 = \frac{c_{Bs}}{2}\left(1-\frac{6c_{Bi}}{c_{Ta}+\frac{c_{Bs}}{2}}+INT\left[\frac{6c_{Bi}}{c_{Ta}+\frac{c_{Bs}}{2}}\right]\right);$$

$$c_3 = c_{Bi};$$

$$c_4 = c_{Ta}\left[\frac{6c_{Bi}}{c_{Ta}+\frac{c_{Bs}}{2}}-INT\left[\frac{6c_{Bi}}{c_{Ta}+\frac{c_{Bs}}{2}}\right]\right];$$

$$c_5 = \frac{c_{Bs}}{2}\left(\frac{6c_{Bi}}{c_{Ta}+\frac{c_{Bs}}{2}}-INT\left[\frac{6c_{Bi}}{c_{Ta}+\frac{c_{Bs}}{2}}\right]\right);$$

And for SSS and ISS with boron in tetrahedral pores:

$$c_1 = c_{Ta}\left(1-\frac{4c_{Bi}}{c_{Ta}+\frac{c_{Bs}}{2}}+INT\left[\frac{4c_{Bi}}{c_{Ta}+\frac{c_{Bs}}{2}}\right]\right);$$

$$c_2 = \frac{c_{Bs}}{2}\left(1-\frac{4c_{Bi}}{c_{Ta}+\frac{c_{Bs}}{2}}+INT\left[\frac{4c_{Bi}}{c_{Ta}+\frac{c_{Bs}}{2}}\right]\right);$$

$$c_3 = c_{Bi};$$

$$c_4 = c_{Ta}\left[\frac{4c_{Bi}}{c_{Ta}+\frac{c_{Bs}}{2}}-INT\left[\frac{4c_{Bi}}{c_{Ta}+\frac{c_{Bs}}{2}}\right]\right];$$

$$c_5 = \frac{c_{Bs}}{2}\left(\frac{4c_{Bi}}{c_{Ta}+\frac{c_{Bs}}{2}}-INT\left[\frac{4c_{Bi}}{c_{Ta}+\frac{c_{Bs}}{2}}\right]\right);$$

Function INT[x] specifies the integer part of x.

Eq. (4) can be used to obtain the expression for average atomic volume V of the solution and calculate distortion energy H_{el}:

$$H_{el} = \frac{1}{2}\left(c_1 B_{Ta}\varepsilon_1^2\overline{V}+c_2 B_B\varepsilon_2^2\overline{V}+c_3 B_B\varepsilon_3^2 Vp+c_4 B_{Ta}\varepsilon_4^2\left(\overline{V}-m\alpha Vp\right)+c_5 B_B\varepsilon_5\varepsilon_5^2\left(\overline{V}-m\alpha Vp\right)\right) \tag{5}$$

In this case, the equation allowing for change in the free Gibbs energy in transition to the solid solution state is as follows:

$$\Delta G = 4c_{Ta}(c_{Bs}+c_{Bi})\Delta H_{mix}^{TaB}+H_{el}-k_B T((c_{Bs}+c_{Bi})\ln(c_{Ta})+c_{Ta}\ln(c_{Bs}+c_{Bi})). \tag{6}$$

To calculate the composition dependences of ΔG for the substitutional and interstitial solid solutions with boron in octahedral and tetrahedral pores, the bulk moduli and atomic volumes of elements were taken from [15] and the mixing enthalpy ΔH_{mix}^{TaB} for the Ta-B equiatomic alloy from [16]. **Figure 12** demonstrates the change in the Gibbs energy for boron-in-tantalum solid solution. The deviation in the lattice parameter a of the solution relative to the lattice parameter a_{Ta} of pure tantalum is determined as

$$\chi = \frac{a}{a_{Ta}} = \sqrt[3]{\frac{\overline{V}}{V_{Ta}}}.$$

The concentration dependence for χ is presented in **Figure 13**. As shown in **Figure 12**, both mechanisms decrease the solution energy. The SSS reaches minimum ΔG at 50 at% B, while the ISS at 33 at% B. Consequently, a combined solid solution may be stable at <50 at% B. The ISS with boron in octahedral pores can be formed only in the presence of the SSS (**Figure 12b**), but the ISS with boron in tetrahedral pores is more likely to appear.

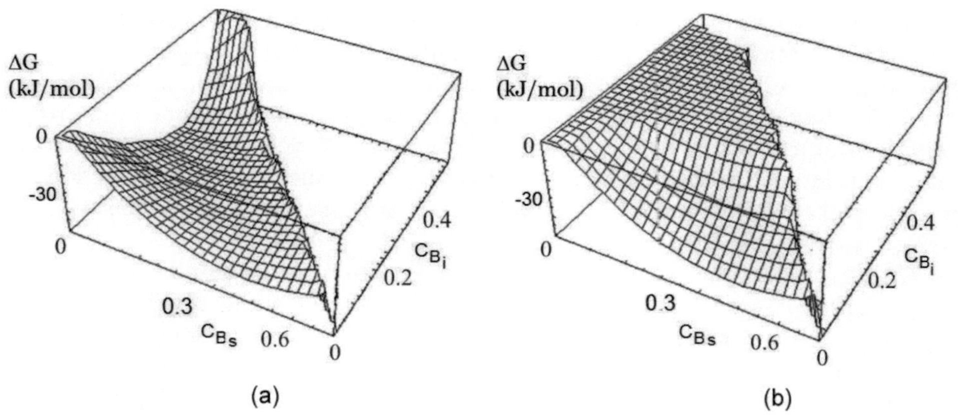

Figure 12.
The calculated Gibbs free energy of solid solutions (ΔG) for combined Ta-B solid solution with boron in tetrahedral (a) and octahedral (b) pores in the bcc lattice of Ta (c_{Bs} and c_{Bi} are the concentrations of substitutional and interstitial boron atoms, respectively).

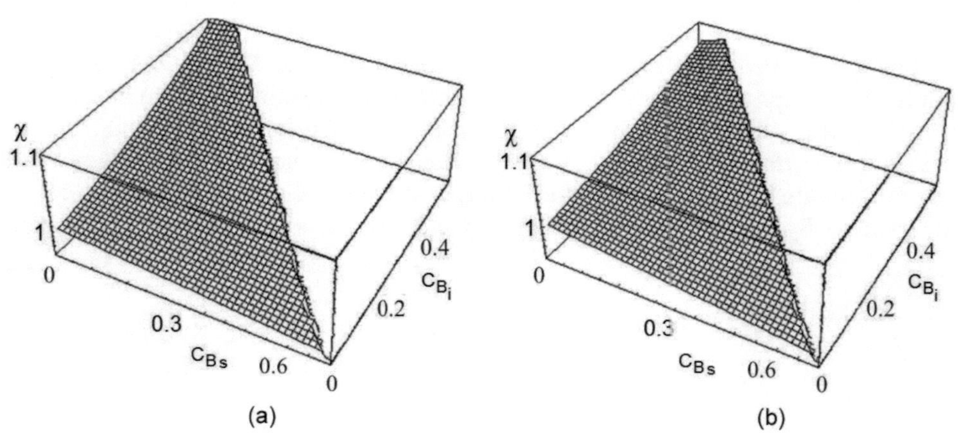

Figure 13.
Change in the parameter χ of the combined (substitutional and interstitial) Ta-B solid solution with boron in tetrahedral (a) and octahedral (b) pores in the bcc lattice of Ta (c_{Bs} and c_{Bi} are the concentrations of substitutional and interstitial boron atoms, respectively).

The parameter χ decreases for the substitutional mechanism and increases for the interstitial mechanism. A greater its increase is observed when boron atoms occupy tetrahedral pores.

3.7 Discussion of results

Modeling for process of formation of boron-in-tantalum solid solutions allows a supposition that the formation of SSS under intense milling of Me-B mixtures in a planetary mill takes place owing to replacement of a metal atom by two (or three for zirconium and hafnium) boron atoms. The possibility of SSS formation through replacement of a metal atom by two boron atoms is due to their close sizes [so-called Goldschmidt criterion (G.c.)]. G.c. was calculated if suppose that two or three boron atoms take place a metal atom knocked out of the crystal lattice under milling according to the formula.

$$G.c. = \frac{2at.v.B - at.v.Me}{at.v.Me} \; x \; 100\% \leq 15\%,$$

where (*at.v.*) is atom volumes for transition metals (*at.v. Me*) and boron (*at.v. B*).

The G.c., atom volumes, and Young moduli for transition metals of IV–VI groups are presented in **Table 1**. G.c. is valid for all of the transition metals except Gr. As for the Young modulus, it is very high for tungsten and molybdenum. That is why those atoms cannot be knocked out of the crystal lattice under milling. Also, a marked difference in the niobium and tantalum Young moduli explains the domination of SSS over ISS in the Nb + 2B mixture in our case. However, under the conditions of less intense milling, ISS prevails [10].

Taking into account the atom volumes for transition metals and boron, one can reveal that replacement of a vanadium atom by two boron atoms results in some increase in the lattice parameter, whereas replacement of a tantalum (as well as niobium and titanium) atom by two boron atoms results in decreasing lattice parameter (**Table 1**). Hagg's rule ($rB/rMe \leq 0.59$) for the ISS formation is fair for tantalum and niobium; therefore, in the Ta-B and Nb-B systems, there are also observed ISS, in addition to SSS. For the V-B system, $rB/rMe = 0.63$, that is, Hagg's rule is not obeyed, so here only SSS can be formed. As the transition metals of group IV Ti and Zr have a *hcp* structure, there cannot be formed ISS. That is why we observed decreasing lattice volume of titanium after 5 min milling with boron. It is min that in this cause only SSS is formed.

In addition, the modeling has showed that at 50 at% B in Ta, the Gibbs energy is minimal. Perhaps at the same boron concentration, an abrupt formation of the TiB_2, VB, VB_2, NbB, NbB_2, and TaB phases occurs due to the minimal SSS stability. The presence of lattice sites replaced by two or three boron atoms in the *bcc* or *hcp* metal lattice facilitates the further transformation of the solid solution into the borides MeB and MeB_2 by cooperative mechanisms under the action of internal strains. This

Group	B	IV			V			VI		
Me		Ti	Zr	Hf	V	Nb	Ta	Cr	Mo	W
Young modulus, GPa		120	98	141	131	105	186	279	325	415
Atomic volume, cm^3/g-atom	4.6	10.6	14.1	13.4	8.3	10.3	10.9	7.3	9.4	9.5
G.c.		13%	2%	+3%	+11%	11%	15%	21%	2%	3%

Table 1.
Young's modulus, atom volumes for boron and transition metals, and Goldschmidt criterion for SSS formation.

is conditioned by the facts that a prototype of chain and ring structures of borides is formed yet in the solid solution, and the significant heat release connected with high enthalpy promotes a reaction running in the regime of mechanically induced self-propagating synthesis. When the fraction of boron atoms incorporated into tetrahedron voids overcomes that taking part in replacement of metal atoms, gradual decomposition of ISS, and the appearance of the TaB_2 take place.

4. Conclusions

The formation of a particular phase under milling in a planetary mill of *hcp* metals (Ti and Zr) and *bcc* metals (V, Nb, and Ta) with boron depends on the physical-chemical properties of starting powders, enthalpy of formation of the final product, and on the intensity of milling process. Depending on these conditions, processes of mechanical alloying in the Me-B system run by the two following mechanisms: mechanically induced reaction of self-propagating synthesis determined by both the enthalpy of refractory compound formation and the capability to form a substitutional solid solution through replacement of a metal atom by boron atoms, and diffusion-controlled process when a supersaturated interstitial solid solution prevails and its *bcc* lattice gradually transforms to the hexagonal lattice of the MeB_2 phase at a critical boron content. The domination of one of the above mechanisms is determined by capability of boron to form substitutional or interstitial solid solution. In the case of formation of combined (ISS and SSS) solid solutions (Ta-B and Nb-B), domination of one of the above mechanisms is determined by the interatomic bond strength, that is, by the Young modulus, as well as by the intensity of milling. The method for calculation the free Gibbs energy of the interstitial and substitutional solid solutions with the regular solution model was developed. It was shown that during milling tantalum and boron powders mixture in a planetary mill, at first the formation of a combined (interstitial and substitutional) solid solution occurs where two boron atoms substitute one tantalum atom. Both the mechanisms of solid solution formation decrease the solution Gibbs energy. When a substitutional solid solution dominates over the formation of an interstitial solid solution, the Gibbs energy acquires a minimum value at a concentration of boron in tantalum of 50 at%, which leads to the solution decomposition and to formation of TaB compound.

Author details

Maria P. Savyak[1]* and Alex B. Melnick[2]

1 Frantsevich Institute for Problems of Materials Sciences, National Academy of Sciences of Ukraine, Kyiv, Ukraine

2 G.V. Kurdyumov Institute for Metal Physics, National Academy of Sciences of Ukraine, Kyiv, Ukraine

*Address all correspondence to: mari_saviak@ukr.net

IntechOpen

© 2020 The Author(s). Licensee IntechOpen. This chapter is distributed under the terms of the Creative Commons Attribution License (http://creativecommons.org/licenses/by/3.0), which permits unrestricted use, distribution, and reproduction in any medium, provided the original work is properly cited.

References

[1] Gild J, Zhang Y, Harrington Y, et al. High-entropy metal diborides: A new class of high-entropy materials and a new type of ultrahigh temperature ceramics. Scientific Reports. 2016;**6**: Article number: 37946

[2] Moiseev GK, Ivanovsky AL. Standard enthalpies of forming related compounds in the metal–boron systems. Izv. Chelyab. Nauch. Ts. 2005; **29**(3):5-9

[3] Suryanarayana C. Mechanical alloying: A novel technique to synthesize advanced materials. Research Official Journal of Cast. 2019. DOI: 10.34133/2019/4219812

[4] Takacs L. Self-sustaining reactions induced by ball milling. Progress in Materials Science. 2002;**47**:355-414

[5] Oghenevweta JE, Wexler D, Calka A. Sequence of phase evolution during mechanically induced self-propagating reaction synthesis of TiB and TiB_2 via magnetically controlled ball milling of titanium and boron powders. Journal of Alloys and Compounds. 2017;**701**(15): 380-391

[6] Jalaly M, Gotor FJ. A new combustion route for synthesis of TaB_2 nanoparticles. Ceramics International. 2018;**44**(1):1142-1146

[7] Gurcan K, Ayas E, Gaşan H. Formation of TaB_2 powders from high energy ball milling and borothermal reduction process. Materials Chemistry and Physics. 2019;**235**:121732. DOI: 10.1016/j.matchemphys.2019.121732

[8] Savyak MP, Melnick AB, Vasil'kivska MA, Timofeeva II, Ivchenko VI, Uvarova IV. Mechanical synthesis of tantalum borides and modeling of solid solutions of boron in tantalum. Powder Metallurgy and Metal Ceramics. 2018;**57**(7–8):373-383

[9] Savyak MP, Melnick AB, Solonin YM, Kotko AV, Timofeeva II, Uvarova IV. Mechanosynthesis of nanodispersed titanium diboride. Powder Metallurgy and Metal Ceramics. 2014;**53**(9–10):497-504

[10] Iizumi K, Sekiya C, Okad S, Kudou K, Shishido T. Mechanochemically assisted preparation of NbB_2 powder. Journal of the European Ceramic Society. 2006;**26**: 635-638

[11] Morris MA, Morris DG. Competition between amorphous or intermetallic phase formation during ball-milling. Journal de Physique Colloques. 1990;**51**(14):151-155

[12] Kiessling R. The borides of tantalum. Acta Chemica Scandinavica. 1949;**3**:603-615

[13] Zhang B, Liao S, Xie H, Yuan X, Shu X. A subregular model for calculating the mixing enthalpies in 10 binary IIB–IIIB alloy systems. Europhysics Letters. 2010;**89**(5): 56002-56016

[14] Takeuchi A, Inoue A. Calculations of mixing enthalpy and mismatch entropy for ternary amorphous alloys. Materials Transactions. 2000;**41**: 1372-1378

[15] Web Elements Periodic Table: The Periodic Table on the Web. Available from: http://www.webelements.com/

[16] Takeuchi A, Inoue A. Classification of bulk metallic glasses by atomic size difference, heat of mixing and period of constituent elements. Materials Transactions. 2005;**46**:2817-2829